Environmental Science, Engineering and Technology

Environmental Effects of Eco-Innovative Coastal Lagoon Dredging for Shoreline Restoration Project in Ghana, West Africa

ENVIRONMENTAL SCIENCE, ENGINEERING AND TECHNOLOGY

Additional books in this series can be found on Nova's website under the Series tab.

Additional E-books in this series can be found on Nova's website under the E-book tab.

ENVIRONMENTAL SCIENCE, ENGINEERING AND TECHNOLOGY

ENVIRONMENTAL EFFECTS OF ECO-INNOVATIVE COASTAL LAGOON DREDGING FOR SHORELINE RESTORATION PROJECT IN GHANA, WEST AFRICA

EMMANUEL LAMPTEY

Nova Science Publishers, Inc.
New York

Copyright © 2011 by Nova Science Publishers, Inc.

All rights reserved. No part of this book may be reproduced, stored in a retrieval system or transmitted in any form or by any means: electronic, electrostatic, magnetic, tape, mechanical photocopying, recording or otherwise without the written permission of the Publisher.

For permission to use material from this book please contact us:
Telephone 631-231-7269; Fax 631-231-8175
Web Site: http://www.novapublishers.com

NOTICE TO THE READER

The Publisher has taken reasonable care in the preparation of this book, but makes no expressed or implied warranty of any kind and assumes no responsibility for any errors or omissions. No liability is assumed for incidental or consequential damages in connection with or arising out of information contained in this book. The Publisher shall not be liable for any special, consequential, or e mplary damages resulting, in whole or in part, from the readers' use of, or reliance upon, this material. Any parts of this book based on government reports are so indicated and copyright is claimed for those parts to the extent applicable to compilations of such works.

Independent verification should be sought for any data, advice or recommendations contained in this book. In addition, no responsibility is assumed by the publisher for any injury and/or damage to persons or property arising from any methods, products, instructions, ideas or otherwise contained in this publication.

This publication is designed to provide accurate and authoritative information with regard to the subject matter covered herein. It is sold with the clear understanding that the Publisher is not engaged in rendering legal or any other professional services. If legal or any other expert assistance is required, the services of a competent person should be sought. FROM A DECLARATION OF PARTICIPANTS JOINTLY ADOPTED BY A COMMITTEE OF THE AMERICAN BAR ASSOCIATION AND A COMMITTEE OF PUBLISHERS.

Additional color graphics may be available in the e-book version of this book.

LIBRARY OF CONGRESS CATALOGING-IN-PUBLICATION DATA

Lamptey, Emmanuel.
  Environmental effects of innovative coastal lagoon dredging for shoreline restoration project in Ghana, West Africa / Emmanuel Lamptey.
    p. cm.
  Includes index.
  ISBN 978-1-61122-140-4 (hardcover)
  1. Dredging--Environmental aspects--Ghana. 2. Lagoon ecology--Ghana. I. Title.
  QH195.G53L36 2010
  577.7'827209667--dc22
                              2010036038

Published by Nova Science Publishers, Inc. † New York

Contents

Preface		vii
Chapter 1	Introduction	1
Chapter 2	Materials and Methods	15
Chapter 3	Results	21
Chapter 4	Conclusion	35
Acknowledgments		46
References		41
Index		55

PREFACE

Evidence suggests that hydraulic dredging is accompanied by considerable adverse environmental impacts in the receiving ecosystem especially on the benthos and water quality. Recently, eco-innovative dredging ('design with nature' principle) is designed not only to minimise environmental impacts but also to enhance the ecological settings. Evaluation of the environmental effects of such eco-innovative dredging is essential to quantify the ecological benefits and the associated impacts, which will ensure sustainable environmental management . Thus, eco-innovative dredging in a large tropical coastal lagoon in Ghana (Keta Lagoon), West Africa, was assessed Before, During and After dredging operations on spatio-temporal scales to ascertain the environmental impacts on the macrobenthic fauna , shorebirds and water quality. A total of 9091 million cubic meters of sediment was removed from an 8 km stretch in the Keta Lagoon for beach nourishment, land reclamation and creation of habitat islands. In quantifying the impacts, macrobenthic fauna samples were collected once in 2000 (Before), 2001 (During) and 2002 (After) along seven stations at 1-km intervals (i.e., A-0 to G-0) in the dredged channel. Water quality was assessed at the subsurface and bottom layers quarterly between June, 2001 and September, 2002, and also in September 2000 before the dredging. Shorebirds' abundance and composition were quantified monthly from August 2000 to 2002, but only parallel data from August-December (peak periods of shorebirds abundance) of each year (2000-2002) was used for statistical analyses.

The results demonstrate that dredging had initial adverse effects on numerical abundance of macrobenthic fauna but there was evidence of community recovery a year after dredging (2002). Species recorded in 2001 (During Dredging) and 2002 (After Dredging) were very similar in terms of composition particularly in the wet periods, suggesting the influence of seasonal environmental factors. The macrobenthic faunal abundance data showed significant spatio-temporal variations ($p<0.05$) and was dominated by opportunistic species of the family Capitellidae. Although, *Nepthys lyrochaeta* showed higher frequency of occurrence (52 %) over the period, there was significant ($p<0.05$) decrease in abundance a year after dredging (2002). Conversely, *Notomastus cf. latericeus* depicted significant ($p<0.05$) increase in population in 2002 (After dredging) possibly suggesting the onset of recovery.

There was no apparent impact on coastal shorebirds although numerical abundance of wader group decreased from 78 % (2000); 69 % (2001) to 51 % (2002). In constrast, terns showed increased abundance from 17 % (2000), 21 % (2001) and 47 % (2002) indicating positive impact. The shorebirds placed in the 'others' category experienced peaks and troughs between the period (6 % in 2000; 10 % in 2001, and 2 % in 2002). In general, mean numerical abundance of the shorebirds increased from 8.8 % in 2000 (Before) to 81.5 % in 2002 (After) revealing an overall strong positive impacts of the project activities on shorebirds.

Temporal and spatial variations occurred in water physicochemical parameters (e.g., salinity, total dissolved solids, total suspended solids and sulfate). Increased turbidity occurred within localized areas during the dredging operation possibly following the direction of the current or the fetch. Nonetheless, the general trends of the water quality variables seem to follow the seasonal pattern in the lagoon. Water quality index indicated that, there was some degree of adverse impact on water quality as the quality decreased from "good" (Before) to "medium" (After).

In all, the results and findings are pertinent to several questions regarding the expected ecological benefits of hydraulic dredging and adverse impacts of the ecology and economy of the receiving ecosystem.

Chapter 1

INTRODUCTION

Coastal lagoons are fragile ecosystems but accounts for 13-15 % of the world's littoral zone [Kjerfve, 1994; Marcovecchio et al., 2005]. They are highly productive than several riverine and estuarine ecosystems in terms of fisheries yield [Kapetsky, 1984], attributable to high primary production as a consequence of nutrients inflow from land drainages [Nixon, 1982]. Coastal lagoons constitute an essential environmental reservoir for fluvial sediments due to their capacity to retain them and also the accompanied elemental pollutants.

In recent times, it has become a common practice to redistribute large amount of coastal sediments for beach nourishment/restoration projects to combat shoreline retreat. The problem of shoreline recession /erosion is as a result of escalating impacts of emerging global geophysical changes, such as rising sea levels, and coastal development [Rakocinski, 1996]. Engineered solutions of massive dredge-and-fill methods have increasingly been adopted worldwide [Bush, 2004; Peterson and Bishop, 2005] in managing the shoreline erosion. But the activities accompanying massive dredge-and-fill projects such as the dredging itself, transportation and disposal of dredged materials result in adverse environmental impacts to the ecology and economy of the receiving ecosystems. Dredging (i.e., underwater excavation of sediment) is carried out for a wide variety of activities and purposes traditionally categorized as (i) capital dredging (e.g., major coastal schemes for a new or expanding port), (ii) maintenance dredging (to maintain the design navigation depth of a waterway) , (iii) remedial dredging (also known as clean-up dredging to improve the environmental quality of a waterway), and (iv) material acquisition dredging (to obtain appropriate sediment material for a restoration project).

There is a huge body of literature that has documented adverse environmental impacts of dredging on benthos, and water quality characteristics [ICES 1992, 2001; Kenny, 1995; Newell et al., 1998; Bemvenuti et al., 2005] but relatively few on aquatic shorebirds [Grippo et al., 2007].

Other reports nonetheless have indicated beneficial uses of dredged materials for beach nourishment, land reclamation, berm creation, replacement fill, capping and habitat creation for wildlife utilization leading to increased biodiversity and economic improvement [see Krause and McDonnel, 2000]. Nonetheless, evidence of benefits of restoration to wildlife is still somewhat ambiguous [Simenstad et al., 2005]. It is uncertain if restoration projects increased habitat use by wildlife or not. Thus assessments of environmental consequences of dredge-and-fill projects are a fundamental step towards understanding the environmental effects particularly of biological resources and their habitats . Despite the huge efforts in monitoring studies, little progress has been made toward understanding and predicting holistic ecological impacts of dredging for shoreline restoration project . Peterson and Bishop [2005] in their review of environmental impacts of beach nourishment pointed a number of areas that need serious considerations in such assessments, including natural temporal and spatial variations in analyses, and rare use of monitoring results to scale mitigation to compensate for injured resources etc.

Relatively very few studies have holistically addressed these issues and also the environmental effects (adverse impacts and beneficial uses) of dredging. Essentially, the extent of dredging impacts on the environment may depend on a number of factors including location, characteristics of the bottom sediment and surface water, the methods of dredging and timing, sensitivity of the ecosystem , and the ecological objectives. The difficulty in predicting the type of response from organisms in ecosystem subjected to dredging requires that studies and analyses of such effects be conducted on a case-by-case basis [Harvey et al., 1998]. Thus, environmental studies investigating impacts of dredging (physical disturbance) on the biological resources and water quality are essential requirements for coastal biodiversity evaluation and subsequent management . Further, when an impact causes complete loss of biological community and deterioration of water quality; studies of environmental impacts are useful in assessing recovery/amelioration and also they constitute an empowering tool for sustainable development.

A variety of study designs existed to detect and assess effects of impacts of general anthropogenic activities in aquatic systems. Green's [1979] BACI

(Before-After-Control Impacts) design have been extensively used in such impact studies where samples are taken before and after a planned impact. It is believed that such design may confound the effects of the impacts with other types of unique natural fluctuations that occur at one site but not at the other [Osenberg and Schmitt, 1996; Hurlbert, 1984; Stewart-Oaten et al., 1986]. This is because anthropogenic sources of stress, often interact with natural processes [Gaston and Edds, 1994; Gaston, 1985; Parker et al., 1980]. A recent design suggests sampling before and after the impacts, at several control sites temporally [Underwood, 1992, 1994].

1.1. GEOMORPHOLOGY AND SEDIMENTATION OF COASTAL LAGOON

A coastal lagoon is a shallow water body separated from the ocean by a barrier, connected at least intermittently to the ocean by one or more restricted inlets, and usually oriented parallel to the shore [Kjerfve, 1994]. All coastal lagoons are recent and transitional geological happenings. The fundamental processes that contributed to their formation took place in the Holocene and were due to changes in (eustatic) sea level about 6000 to 7000 years ago [Emery and Stevenson, 1957; Phleger, 1969] together with the rise and fall of the coastal area [Zenkovitch, 1969]. In the case of an emergent coastline, the shallowness of the water and the plain may give rise to a submerged beach that contributes to the formation of a barrier, which isolates interior water and, thus, forms a lagoon. In the case of sinking coastline, the mechanism that operates is a continual and gradual rise in sea level forming a barrier [Zenkovitch, 1969] or with regard to a slight sloping coast, the coastal water from the sea may flood a coastal depression , in a process that under certain circumstances may form a barrier that encloses the depressions and forms a lagoon [Lankford, 1977].

In geomorphologic terms, coastal lagoons usually occur where valley mouths or lowlands have been submerged by the sea during the later stages of the Late Quaternary marine transgression , which on tectonically stable coasts brought the sea up to approximately its present level about 6000 years ago [Bird, 1994]. Once formed, coastal lagoons are modified by erosion and deposition . Infilling by accumulation of in-washed sediment , organic deposits such as peat or shells, precipitated material, results in shallowing and shrinkage of lagoons [Bird, 1994]. Contrasting sea-level histories have e rted a fundamental control on coastal sedimentation [Dominguez, 1984].

Under conditions of sea-level rise, Barrier Island/lagoonal systems become important environments of sedimentation. Barrier islands form preferentially under conditions of sea-level rise. According to Bruun's rule [Bruun, 1962], if the coastal profile is accepted as an equilibrium response of the sea floor to the coastal fluid power expenditure, then the effects of a sea-level rise could be deduced as a landward and upward translation of the profile. Thus, as sea level rises along a low-relief coastal plain, the beach and dune are nourished by the longshore drift and grow upwards at the same rate of sea-level rise following Bruun's rule. The swale behind the dune, however, remains at the same altitude and, as sea level rises, becomes a lagoon [Martin and Dominguez, 1994]. When sea level falls, the inverse of Bruun's rule applies, resulting in a seaward and downward translation of the coastal profile, and shallow back-barrier lagoons eventually become emergent [Martin and Dominguez, 1994]. Because lagoons are generally shallow, they are very sensitive to fluctuations of sea level, small rises and falls translate respectively into widespread inundation and emergence of coastal lagoons. Thus, coastal lagoons during their geological history may be affected by multiple episodes of invasion and emergence.

Ecological conditions, particularly water salinity and temperature, are important in the geomorphological evolution of coastal lagoons. They control the extent to which vegetation can colonize lagoon shores, impeding erosion, promoting pattern of sedimentation and generating organic deposits [Bird, 1994]. Coastal lagoons around the world show great contrasts, but the same processes have operated in similar situations. In general, coastal lagoons are formed and maintained through sediment transport processes. Sediments are carried by rivers, waves, currents, wind, and tides [Nichols and Boons, 1994]. Lagoons fed by rivers receive sediments ranging from coarse sand to silt and clay. The coarser material is deposited as the river enters the lagoon, and may be added to lagoon beaches and spread around the shore by wave action; the finer sediment is carried out into the lagoon and deposited on the floor, progressively reducing the depth. Rates of fluvial sediment yield to lagoons may be accelerated by the reduction of vegetation cover and the onset of soil erosion in the river catchments [Bird, 1994]. Studies have also indicated that higher energy conditions are responsible for the composition and distribution of sediments in a lagoon [Hubbard, 1992; Kalbfleisch and Jone, 1998; Kench, 1998]. The development of these sediments is attributed to short-lived, storm-induced high energy conditions [Beanish and Jone, 2002].

1.2. VALUE OF COASTAL LAGOONS

Coastal lagoons present many ecological and socio-economic values, albeit the debate as to whether natural systems usurp an intrinsic value of their own that lies outside human determination is waging [Rolston, 1994; Williams, 1994]. Coastal lagoons support a wide range of natural services that are highly valued by society including fisheries, storm protection and tourism. They are important as nursery grounds for a variety of marine fishes and shrimps [Day and Yanez-Arancibia, 1985; De Wit, 2003]. Significant fisheries of oyster, shrimp and bony fish exist within many lagoons. Migrating birds make extensive use of lagoons, where they feed, roost and may spend most, or all of their lives there (Armah, 1993). Lagoons serve as sanctuaries in certain areas for endangered species such as crocodiles and hippopotami [Day and Yanez-Arancibia, 1985]. Many of the coastal lagoons serve as important harbors, navigation routes and also for recreation. They therefore, constitute important sites for industries that are connected with tourism. The shores of certain coastal lagoons are preferred locations for construction of residences owing to its numerous benefits.

Lagoon reefs made of molluscs are extensively used for concrete aggregate in low coastal areas, which are far from source hard rock. Some lagoons are dumping areas for disposal of waste from urban and industrial areas. They constitute areas for the production of significant amount of salt from local techniques of evaporated pans (e.g., Keta and Songor lagoons in Ghana). In some places, their water is used for the cooling of generators of electric power plants, which return effluent of warmed water to the lagoon. Coastal lagoons are economically important in their use for aquaculture facilities [Day and Yanez-Arancibia, 1985; De Wit, 2003]. Coastal lagoons facilitate processes, particularly those resulting in loss or accretion of natural wetland. They provide numerous ecosystem goods and services.

1.3. ENVIRONMENTAL IMPACT STUDIES

The *environment* is the sum total of all causal factors that show actual interactions. It comprises the input and output components, with resources and conditions constituting the input environment. Environmental conditions are all things outside an organism that affect it but, in contrast to resources, are not consumed by it [Begon et al., 1990]. *Impact* is defined as any effect caused by human activity (activity is basic element of a project that has potential to affect any aspect of the environment) on the environment including flora, fauna, sediment, air, water, climate, landscape or other physical structures or the interactions among these factors [Espoo, 1991]. An *environmental impact* is an estimate or judgement of the significance and value of environmental effects on physical, biological, social or economic environment (Espoo, 1991).

Environmental Impact Assessments (EIAs) is a requirement for any project that may have the potential to cause significant impacts to the environment. A principal objective of the EIA (Ghana perspective) is to provide enough relevant information to the Environmental Protection Agency (EPA) to enable the Agency to set an appropriate level of assessment for a proposed project. The information collected through the EIA may be presented in one of several forms, but for larger projects, the most common is an Environmental Impact Statement (EIS).

Environmental impact studies are composed of two distinct phases (i) impacts analysis phase, which is meant to identify, predict, quantify and evaluate the effects of expected impacts before a project occurs and (ii) a monitoring and assessment phase, which is meant to measure and interpret environmental effects during the project and after it has been completed. Impact hypothesis draws on the results of earlier studies of environmental characteristics and their variability [British Geological Survey, 1999]. Impacts whether they are significant or not, can be direct or indirect.

Direct impacts to biological resources result when biological resources or critical habitats are altered, destroyed, or removed during the course of project implementation. Indirect impacts to biological resources may occur when project activities result in *environmental change* (i.e., measurable change in physical and biological systems and environmental quality resulting from a development activity) that directly influence the survival, distribution, or abundance of native species (or increase the abundance of undesired nonnative species). It is also possible to have beneficial impacts, directly or indirectly. Impacts may also be short- or long term. Short-term impacts are generally not considered significant, by definition. Impact thresholds are based on factual evidence of physical disturbance of habitat, and the loss or disturbance of recorded species. Impact thresholds could have significant impacts on biological resources (i.e., benthos and shorebirds) through (i) loss in substantial number of individual of species or loss that could affect abundance, (ii) a substantial adverse effect on species, natural community, or habitat which is recognised as biologically significant, (iii) significant degradation of pelagic or benthic habitats for rare species and/or native species, and (iv) disruption of the trophic structure of the biological communities.

Biological communities exhibit complex interacting behaviours among themselves and with the non-living abiotic environment [Lamptey and Armah, 2008]. These communities play multiple ecological roles within an ecosystem and therefore, are a critical part in monitoring and evaluation of project impacts.

In most environmental impacts studies, benthic invertebrates are the principal targeted organisms (78 percent of all studies), reflecting their suitability as ecological indicators [Peterson and Bishop, 2005]. These organisms are operationally classified as microbenthos (< 63 µm), meiobenthos (from 63 µm to 500 µm) and macrobenthos (> 500 µm or > 1000 µm) according to the sieve mesh size used for extraction from sediment cores or grabs. The macrobenthos forms are by far the better known and used component of the marine biota in environmental impact studies [Clarke and Warwick, 1994]. The macrobenthos, composed mainly of molluscs (shellfish and snails), polychaetes (bristle worms), crustaceans (amphipods, shrimps, and crabs) and echinoderms (sea cucumbers and brittle stars) [Gray, 1981], is the infaunal component most widely used in environmental impact studies [Pearson and Rosenberg, 1978; Pocklington and Wells, 1992; Paiva, 2001]. The potential benefits of using macro-invertebrates include quick detection of pollution through differences between predicted and actual faunal assemblages [Ormerod and Edwards, 1987]. Benthic invertebrates are relatively sessile (therefore allowing spatial patterns to imply causation), can be sampled quantitatively without high cost, are well described taxonomically, and reveal ecologically meaningful and important patterns, even at coarse levels of taxonomic discrimination [Warwick, 1988]. Thus, in assessing the impacts on biological resources, the temporal and spatial variability of the benthic assemblages along with predicted area and rate of recolonization constitute a good consideration. Changes in soft bottom zoobenthic communities in response to the environmental impact have been successfully implemented world-wide in pollution assessment studies and monitoring programs [Pearson and Rosenberg, 1978].

1.4. ENVIRONMENTAL IMPACTS OF DREDGING

The principal biological impacts of dredging include disturbance and removal of benthos and alteration of the substrate upon which benthic colonization depends [British Geological Survey, 1999]. The environmental impactss of dredging have been well documented, with general reviews of the topic provided by ICES [1992, 2001]; Kenny et al. [1998] and Newell et al. [1998]. It was clear from such reviews that most studies have been concerned with impacts of dredging on soft sediments or those associated with beach nourishment projects. A direct impact of dredging would come from loss of invertebrates through mortality and removal of sediment . Physical removal of substratum and associated plants and animals from the seabed, and burial due to subsequent deposition of material are the most likely direct effects of dredging and reclamation projects [Newell et al., 1998]. New habitats may also be created as a result of the operation, either directly in the dredged area or by introduction of new habitats on the slopes of a reclaimed area (e.g. hard substratum in the form of breakwaters and revetments).

Dredged material may come into suspension during dredging itself as a result of disturbance of the substratum, and also during transport to the surface, overflow from barges or leakage of pipelines, during transport between dredging and disposal sites, and during disposal of dredged material [Jensen and Mogensen, 2000]. Dredging may change the physical environment and could directly impact on macrobenthic organisms through (i) compaction of sediment , (ii) burial of organisms, or (iii) smothering through increased turbidity and siltation [Goldberg, 1989]. Dredging has effects at two locations, the site of removal and the site where the material is dumped [Hall, 1994]. The natural processes of sedimentation in coastal lagoons are significantly altered by dredging activities. The consequence changes in sediment composition as a result, affect the associated macrobenthic fauna . Bonsdorf [1983] examined re-colonization after dredging at three shallow brackish sites in Finland . The study showed that the pool of available colonists is important in determining the dynamics of disturbed patches. At one site, dredging occurred to below the thermocline and the benthos at this level were exposed to deoxygenation events every year, which defaunated the sediment. Deoxygenation took approximately two months to kill the fauna and this was followed by a gradual recovery with the peak in species richness occurring after about 10 months.

In contrast, just above the thermocline, a stable community developed over the six years of the study and this region provided colonists for deeper parts. With the progressive recovery of the upper region, a more diverse and abundant pool of colonists was available to recolonize the deeper parts, which led to successively higher peaks in species richness each year. At a second site at 8-9 m depth in a channel, it took 4-5 years for the community to return to a background level, despite the area containing only about three species. Interestingly, early in the colonization sequence, three species established which had not occurred in the area before dredging. Species richness declined after five years and these three species were not found in the final community which itself contained only three species.

Maurer et al. [1986] reviewed studies of burial effects and concluded that the pattern of susceptibilities can be reversed when sediments containing silt/clay are compared with those comprising sand. This was based on earlier experimental studies, which indicated that atypical sediments for the area caused the highest mortalities in estuarine bivalve species following burial in natural and exotic sediments [Maurer et al., 1986]. Maurer et al. [1986] cited Kranz [1972] who studied the burrowing of 30 species of bivalves showed that the life habits of the taxa affected the susceptibility of the fauna to mortality. Mucous tube feeders and labial palp deposit feeders were most susceptible, followed by epifaunal suspension feeders, boring species and deep burrowing siphonate suspension feeders, none of which could cope with more than 1 cm of sediment overburden. Infaunal non-siphonate suspension feeders were able to escape 5 cm of their native sediment, but normally less than 10 cm. One potentially complicating factor, when considering the effects of dumping dredge spoil, is that many types of sediment will be contaminated [Hall, 1994]. Indeed, much of the motivation for studies on dumping dredge spoil effects stems from concern over chemical pollutants rather than dumping per se.

Flemer et al. [1997] concluded that there was no apparent consistent gross effects of dredged material disposal on macrobenthic community structure at coastal Louisiana and suggested that some long-term unidentified factor (e.g. sediment toxicity) maintained differences in macrobenthic community structure in the three different study areas. A number of factors influence the effects of dredging on local populations including the types of organisms which remain in the vicinity [Thrust et al., 1992], the life histories and mechanisms of dispersal in different fauna [Levin, 1984], the patchiness of the environment [Hall et al., 1994], the spatial and temporal variability of dredging disturbance, the effects of existing or new residents on the substratum [Rhoads and Boyer, 1982; Davoult and Richard, 1990] and the potential interactions between dredging disturbance and other perturbations [Jewett et al., 1999].

Dredging activities are also usually associated with profound changes in water quality as a result of alterations to natural suspended sediment loads and localized sedimentation resulting in clouding and colouring of the surface water. Impacts associated with increased suspended particles in the water column include high turbidity levels, reduced light transmittance and reduction or loss of benthic habitats. The intensity and duration of sediment re-suspension from dredging and disposal operations is highly dependent on the type of equipment, operator, characteristics of sediment, and the hydrodynamic conditions [Collins 1995; Clark and Wilber, 2000]. Elevated suspended sediments have also been shown to adversely affect the respiration of fishes, reduce egg buoyancy, disrupt icthyoplankton development and reduce filtering activities of benthic organisms [Messieh et al., 1991; Barr, 1993].

Other forms of dredging like aggregate mining can reduce localized current strength, resulting in lowered dissolved oxygen concentrations. Reduced oxygen levels adversely affect the ability of fish and invertebrates to utilize specific areas for spawning, feeding and development [Pacheco, 1984]. The release of material into the water column during dredging can alter water quality, especially if excavated material is high in organic matter. The effects of mixing in the water column are likely to increase the demand of oxygen by decomposing organic matter and the release of nutrients [ICES, 1992]. Furthermore, dredging could increase or decrease the exchange rate of nutrients between the sediments and water column and introduce pulses of productivity during nutrient recycling. Dredging activities therefore affect certain physical and chemical conditions of the water such as degree of oxygenation and mineralization, temperature, salinity, water flow, depth and water level fluctuations.

The actual impact of dredging operation will be a function of the spatial extent and degree to which the post-construction environment differs from pre-construction conditions.

1.5. OVERVIEW OF THE KETA SHORELINE RESTORATION PROJECT

The Keta Lagoon is an ecologically important wetland in West Africa and has been recognized as a designated Ramsar site (a wetland of international importance especially as a habitat for waterbirds). Ghana has ratified both the Ramsar and Bonn (which addresses the conservation of migratory species of wild animals) Conventions in 1988.

The Keta Lagoon covers an estimated area of 340 km^2 with water depths ranging from 0.47 to 0.94 m in the wet season and 0.14 to 0.56 m in the dry season [Lamptey and Armah, 2008]. The lagoon has a maximum coastal length (east–west) and width (north–south) of 25 and 13.5 km, respectively [Lamptey and Armah, 2008]. It is separated from the sea by a narrow sand bar (Fig. 1) and, therefore, receives sea water only through spillover during periods of high tide. The Keta basin was formed by coastal subsidence during the Precambrian [Akpati, 1975]. The upper geologic strata (about 24 m) are composed of coarse, unconsolidated beach sand and gravels both of fluviatile and shallow marine to estuarine origin [Akpati, 1975]. Most areas in the lagoon are typically muddy in the upper 10 cm. The sea grass *Ruppia maritima* used to occur in the northeastern part of the lagoon and portions of the southern part until it disappeared in 2004. The macrophytic flora in the lagoon is dominated by *Typha domingensis* and *Paspalum vaginatum* in the northwestern and southwestern portions in the freshwater tributaries of the lagoon. The southwestern part is dominated by *Paspalum vaginatum*.

The lagoon receives fresh water from a large catchment area including (1) runoff from the Tordzie River, which originates from the Akwapim–Togo ranges; (2) runoff from the Aka and Belikpa catchments, which enters the lagoon from the north; and (3) inflows from the Volta estuary through Anyanui creek [Entsua-Mensah and Dankwa, 1997]. The Tordzie River has a catchment area of 2,200 km^2 and a mean annual flow of 11 m^3 s^{-1}; Aka and Belikpa have catchment areas of 280 and 420 km^2, respectively; the total drainage area of the Volta estuary is 37,900 km^2 [Finlayson et al., 2000]. Nevertheless, the volume of water (84,446 m^3) transferred to the lagoon during one flood period in January 2001 from the Volta estuary via Anyanui creek resulted in a tidal excursion of 5.4 km [Sørensen et al., 2003], indicating that the fresh water that flows from the Volta estuary into the lagoon is not substantial. The estimated static capacity of Keta Lagoon is 360×106 m^3 when there is no flow of water into it [Finlayson et al., 2000]. The area lies within the dry equatorial region of Ghana, which has two wet seasons, one from May to July (major rainy season) and from September to November (minor rainy season). The mean annual rainfall is 750 mm [Dickson and Benneh, 1988]. The dry season begins in January and ends in March. Annual mean air temperatures range between 24° C and 32° C. Evaporation in the area far exceeds annual rainfall. It is only during the major wet season that monthly rainfall may exceed evapotranspiration and temporary streams flow [Biney, 1986]. The prevailing wind direction is from the southwest (the southwest monsoon), which is a feature of the entire coastal belt of the country [Finlayson et al., 2000]. The mean monthly averages of daily wind speeds range from 5.86 to 8.06 m s^{-1} [Finlayson et al., 2000].

Restoration project of the Ketaarea was occasioned by previous episodes of erosion threats dated back in 1920s. The Keta sea defensedefense project began in the year 2000 by Great Lakes Dredge and Dock (GLDD)after a detailed study in 1996, which recommended an engineering solution [ESL/RPI/GLDD, 2004]. Prior to the restoration project, flooding from the Keta Lagoon following torrential rains was a perennial eventthat has displaced many of the coastal communities in the area. The project had four principal components: i) sea defense, ii) land reclamation via beach nourishment; iii) construction of a road along the lagoon; and iv) construction of a flood control/relief structure (Fig.1). As a complement to the specific objectives of each component, an overall ecological consideration of the project was to minimize impact to, or even improve the general environmental conditions and enhance the ecological setting against the background of a Ramsar site. Restoration science, and its recent manifestation, sustainability science are partly rooted in ecological fidelity [Kates et al., 2001], i.e., those restoration goals are characterized by structural replication, functional success, and durability (self-sustainability) [Higgs, 1997].

In order to obtain appropriate sediment materials for the road construction, beach nourishment, and infilling the reclaimable land, 8.5 km stretch of the Keta Lagoon was dredged and approximately 9, 091,000 m^3 of sediment was removed. However, aware of the adverse impacts associated with dredging and the ecological importance of the lagoon in supporting internationally important bird species ,an eco-innovative dredging was designed to use the dredge spoil to create habitat islands in the lagoon (Fig. 2). As a consequence of the project , a total area of approximately 283 ha of land was reclaimed by construction of groynes from the shore to a depth of approximately 4.5 meters NLD (National Land Datum) and extending 140–200 meters from the existing shore at the seaward section (Fig. 3).

1.6. OBJECTIVES

The overall objective of this study was to investigate the environmental impacts of eco-innovative dredging on the receiving ecosystem on spatial and temporal scales. The specific objectives were: (i) to assess the impacts of the dredging on macrobenthic fauna spatio-temporally, (ii) to ascertain the effects of the dredging activities on shorebirds' abundance, and (ii) to determine the dredging effects of the water quality of the lagoon.

Introduction

Figure 1. Flood control/relief structure and view of 8-km road stretch across southeastern portions of the lagoon.

Figure 2. View of habitat islands created for shorebirds using dredged material.

Figure 3. Aerial photograph of embayments between groynes, nourished area, settlements, dredged channel, and habitats island.

Source: GLDD/ESL/RPI.

Chapter 2

MATERIALS AND METHODS

2.0. FIELD SAMPLING

The sampling was designed to provide a sound monitoring plan that assessed the environmental impacts of the eco-innovative dredging in the Keta Lagoon ecology . The sampling design followed sample collection before, during and after the dredging activities. Following the design seven impacted stations were located of 1-km intervals in the dredged channel. The stations were labelled alpha-numerically (A-0 to G-0) (Fig. 4).

Quantitative sampling of macrobenthic fauna was carried out using an Orange-peel grab at each of the stations in September, 2000, 2001, and 2002 before, during and after dredging respectively. Four replicate samples were taken at each site. A grab sample constituted a sediment volume of 2.036×10^{-3} m^3. The samples were processed by passing through a 0.5 mm sieve (to retain macrobenthic fauna) and fixed with 10 % borax pre-buffered formaldehyde solution for later taxonomic resolution in the laboratory.

Water quality parameters were collected quarterly at the surface and bottom (During and After dredging) using a Van Dorn water sampler. Nonetheless, water samples were also collected at subsurface on the stations before the dredging operations. In-situ measurements of dissolved oxygen, temperature , pH , water depth, salinity , transparency were carried out at each of the locations. Additional water samples were collected for laboratory analyses of nutrients (nitrate, phosphate, silicate, and sulfate), turbidity, suspended solids, and dissolved solids.

Shorebirds observations were made using a 60 mm, 15-60x zoom spotting scope mounted on a tripod and 7 x 50 binoculars. Observations of the shorebirds were made in the morning (6:00 AM). Five monitoring stations within the sphere of the project area were visited in each survey session to quantify abundance and composition of the shorebirds.

2.1. LABORATORY SAMPLE PROCESSING

Macrobenthic faunal samples were washed thoroughly with fresh water in 450-μm stainless sieves to get rid of the formaldehyde solution and excess mud. The washed sediment samples were separately placed in 50 x 40 m sorting tray with a white background thinly spread out with small amount of water covering it. The organisms were picked into storage vials with the aid of hand lens and preserved in 70 % ethanol mi d with glycerol (glycerol reduces the evaporation of the ethynol) . During sorting the organisms were grouped into broad taxonomic units such as polychaetes, mollusks , crustaceans etc. These broad taxa were identified to genus or species levels as possible and counted. Identifications were based on taxonomic guides and manuals [e.g., Day, 1967a, b; Edmund, 1978] as well as voucher specimens in the Zoological Museum of the University of Ghana. Voucher specimens are available for examination.

Water samples collected at the subsurface and bottom were analyzed in the laboratory for nitrate, phosphate, silicate, sulfate, conductivity , turbidity, suspended solids, and dissolved solids using the HACH DR/2010 spectrophotometer following the methods in A.P.H.A. et al. [1998].

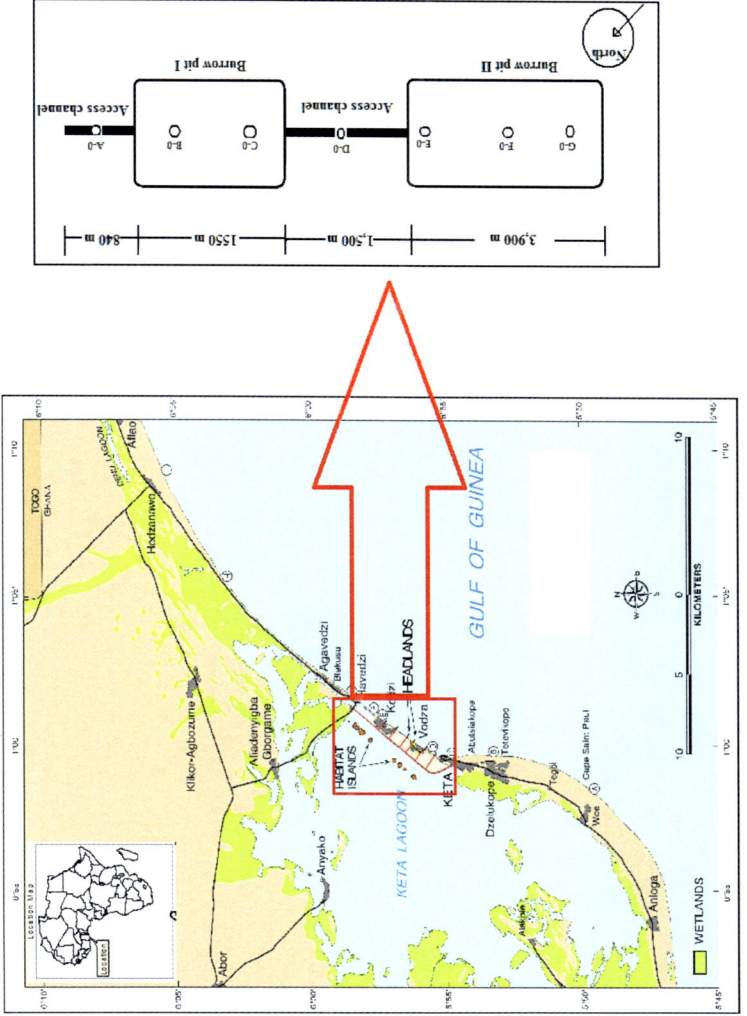

Figure 4. Map of Keta Lagoon showing study area, schematic diagram of relative locations of burrow pits and access channels and their approximate lengths as well as sampling stations of the dredged channel.

2.2. DATA ANALYSES

The data set was analyzed using both univariate and multivariate statistics. Spatial and temporal distributional trends were plotted to indicate the impacts of the project on the macrobenthic fauna, shorebirds and water quality.

To test for differences in macrobenthic faunal community structure between the periods; one-way Analysis of Similarity (ANOSIM) was utilized [Clark and Warwick, 1994]. This program computes r-statistics as a measure of discrimination. First, a global R-value was computed to indicate the overall effect of similarity between the study shores. Values of R=1 are obtained when all replicates (stations) within the groups (year) are more similar. The p-value for the statistics was obtained by simulating all possible permutations of assigning replicates (stations) to study years. In this study, a random sample of 999 permutations was used in each calculation.

The species that contributed the most and discriminated one year from another were investigated using non-metric similarity percentage procedure (SIMPER) [Clark and Warwick, 1994]. These results assisted in the interpretation of the community changes responsible for the observed pattern [Clarke, 1993]. The group of species with cumulative contribution above the 50 % (dis)similarity threshold were considered important in controlling the taxa assemblages in the study area.

Water quality index (WQI) was calculated to determine the deterioration of quality or otherwise as a result of the dredging activities. For calculating WQI, the following steps were used:

In the first step, unit weight (Wi) for various parameters is inversely proportional to the recommended standard ($V_{standard}$) for the corresponding parameter. Wi values were calculated by using the following formula:

$iW = K/V_{standard}$

where, K = proportionality constant, $V_{standard}$ = world-widely accepted drinking water quality standard prescribed by the World Health Organization (WHO) [WHO, 2004]. The constant of proportionality K in the above equation can be determined from the following condition,

$iW = \sum K (1/V_{standard})$

In the second step, Quality rating (Qi) is calculated as,

$Qi = 100 \ (V_{actual}/V_{standard})$

where, V_{actual} = value of the water quality parameter obtained from the laboratory analysis.

$V_{standard}$ = value of the water quality parameter obtained from recommended WHO standard of corresponding parameter.

This equation ensures that $Qi = 0$ when a pollutant is totally absent in the water sample and $Qi = 100$ when the value of this parameter is just equal to its permissible value. Thus the higher the value of Qi is, the more polluted is the water.

Then, the overall WQI was calculated on the basis of weighting and rating of the different physico-chemical parameters, as follows:

$WQI = \sum W_i Q_i$

The resultant values for period was substracted from 100 to obtain the WQI based on the index scale. In other instances, the calculated values may not be substracted from 100 depending on the water quality interpretation scale. The water quality variables used in the calculation of the WQI were pH, dissolved oxygen, turbidity, total dissolved solids, nitrate and phosphate.

Chapter 3

RESULTS

3.1. DISTRIBUTION OF DREDGED MATERIALS ON RESTORATION SITES

Two hydraulic cutter dredges were used to remove varied quantities of sediments in two burrow pits and access channel of the 8 km stretch of the Keta Lagoon (Fig. 4) to obtain sediments for the restoration. The average depths of the burrow pits were approximately 9.6 m and 11.1 m for pits 1 and 2 respectively (Fig. 4). The depths for the access channels ranged between 2-4 m. The burrow pits were approximately 300 m wide while the access channels were 40 m wide. Table 1 presents the quantities of sedimentary materials deposited at each construction site. The total sedimentary material deposited on each constructional site varied. Approximately, 2,861 million cubic meters dredged sedimentary material was deposited for beach nourishment (see Fig. 3) with a daily average production of approximately 11,800 cubic meters. The total dredged material production to the habitat islands was 2,010 million cubic meters with average daily production of approximately 9,050 cubic meters. At the site for reclamation, a total of 4,220 million cubic meters and a daily production of approximately 10,300 m^3 were deposited. The water to solids ratios of the sediment materials were 10–16 % in medium sand and 12–20 % in cohesive soils [ESL/RPI/GLDD, 2001-2004]. The dredging activities in the lagoon varied throughout the period according to actual sand requirements , but it was estimated that on average 300,000-350,000 m^3 of sediment were dredged monthly.

Table 1. Average and total dredged sediments deposited at the project specific area

Project Component	Average daily Production (m^3)	Total material deposited (m^3)	No. of days for deposition
Beach Restoration	11,800	2,861,000	242
Land Reclamation	10,300	4,220,000	410
Habitat Island	9,050	2,010,000	222
Total	31,150	9,091,000	874

3.2. IMPACTS OF DREDGING ON SHOREBIRDS' ABUNDANCE AND COMPOSITION

The total abundance of shorebirds increased considerably from 12,346 in the year 2000 (Before dredging), 13,643 in 2001 (During dredging) to 114, 398 in 2002 (After dredging). The mean numerical abundance of shorebirds showed significant ($p<0.05$) increase after dredging in 2002 (Fig 7) suggesting overall positive impact of the dredging activities on shorebirds' abundance. Nonetheless, the numerical abundance of 'wader birds ' group decreased after the dredging activities in 2002 (Fig. 8) indicating negative impact. Conversely, 'terns' category showed increased trend from the year 2000 to 2002 due to the creation of the habitat island (Fig.5), whereas the 'others' category fluctuated between the periods (Fig. 8).

Figure 5. Terns on one of the Habitat Island in the Keta Lagoon, April 2001 (Photo A.K. Armah).

Figure 6. Nesting and fledgling of Little tern (*Sterna albifrons*) on habitat island in the Keta Lagoon, August 2000 (Photos L. Cotsapas).

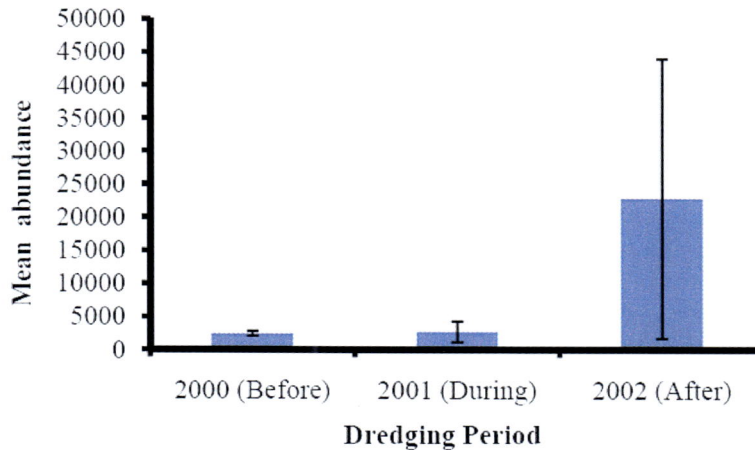

Figure 7. Mean total abundance of shorebirds Before, During and After the dredging period (Abundance was computed from August to December of each year).

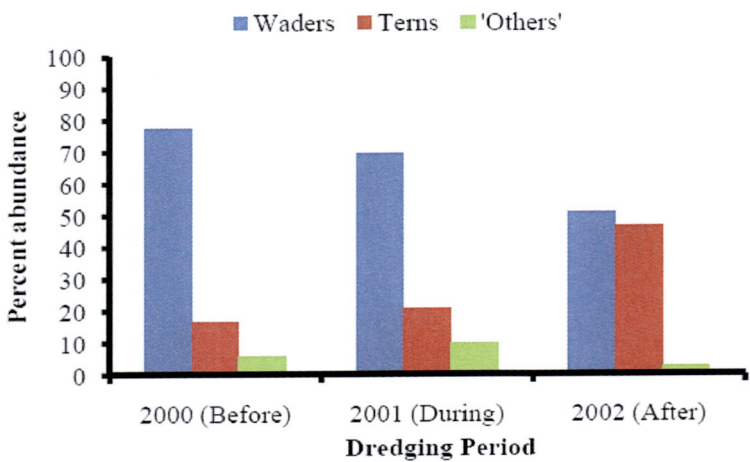

Figure 8. Percent abundance of major shorebird groups 'Before', 'During' and 'After' dredging (Abundance data was computed from August to December of each year).

3.3. DREDGING IMPACTS ON MACROBENTHIC FAUNAL COMMUNITY STRUCTURE

The macrobenthic faunal community of the dredged portion of the lagoon showed significant reduction in abundance (using one-way ANOSIM) between the year 2000 (Before) and 2001 (During) (r=0.205, $p<0.05$), as well as 2000 (Before) and 2002 (After) (r=0.166, $p<0.05$) of the dredging periods. The species that contributed to the significant difference between the periods were *Ischyroceros* sp., *Nereis operta*, *Notomastus* cf. *latericeus*, *Nephtys lyrochaeta*, *Tellina nymphalis*, *Capitella* spp. (Table 2). The first four species contributed greater than 50 % of the average dissimilarities of 85.29 % and 83.57 % realized respectively between 2000 and 2001, and also between 2000 and 2002. However, three species namely: *Nepthys lyrocheata*, *Tellina nymphalis* and *Notomastus* cf. *latericeus* contributed greater than 50 % to the average dissimilarity of 87.71 % between 2001 and 2002. The differing abundances of these species largely influenced the macrobenthic faunal assemblage structure of the dredged channel.

Individual macrobenthic faunal species showed both spatial and temporal differences (Figs. 10 & 11) in numerical abundance such that higher numbers occurred at Stations C-0 (*Tellina nymphalis*) and G-0 (*Nephtys lyrochaeta*, *Notomastus* cf. *latericeus* and *Capitella* spp.) in 2001. This could possibly due to the fact that Stations F-0 and G-0 were freshly dredged during the sampling period. A notable feature was the absence of *mactra nitida* at any of the stations before dredging but occurred in appreciable numbers in 2001 (Fig. 11) due perhaps to condusive conditions. *Capitella* spp. was abundant at Station D-0 in 2002 and also significantly ($p<0.05$) hgher abundances in 2001 and 2002 compared to 2000 (Fig. 11) indicating some level of habitat disturbance. This is because *Capitella* spp. are known to be opportunistic and exist in stressful habitats .

The effect of spatial declension in species abundance after dredging (2002) was significantly pronounced at Stations A-0, B-0, E-0, F-0 and G-0 (Fig. 10). Conversely, Stations C-0 showed significant increase in both species abundance and number of species after dredging in 2002 (Figs. 10 & 12). In the same vein , Station D-0 depicted increased abundance and evidence of community recovery (Figs. 10 & 12). Certain stations notably Stations A-0, B-0, E-0 and F-0 recorded no species after dredging. Probably, the dredged material mainly peat smothered and killed the species.

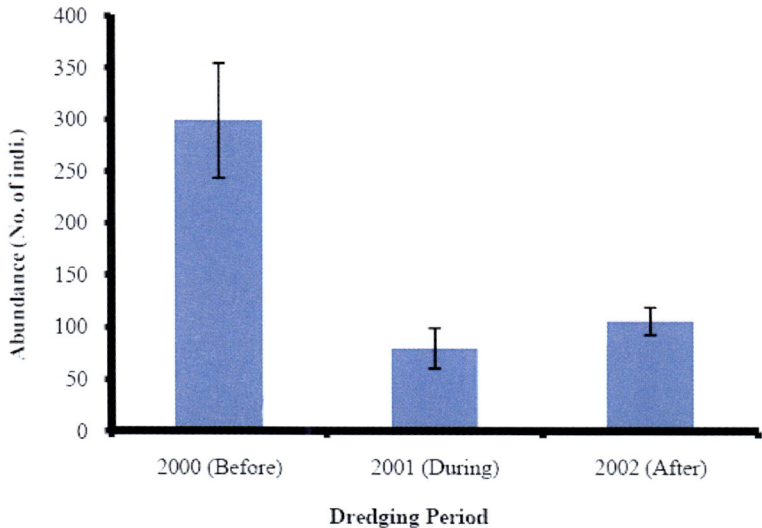

Figure 9. Temporal distribution of macrobenthic faunal abundance before, during and after dredging. Error bars indicate 95% confidence interval.

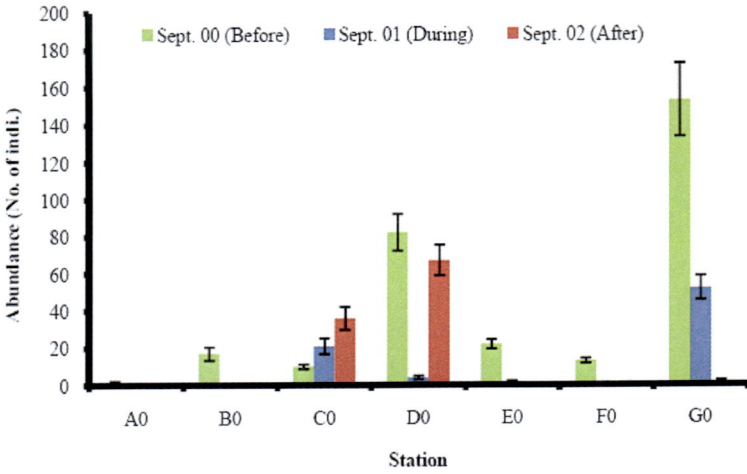

Figure 10. Spatial abundance of macrobenthic fauna before, during and after dredging. Error bars indicate 95% confidence interval.

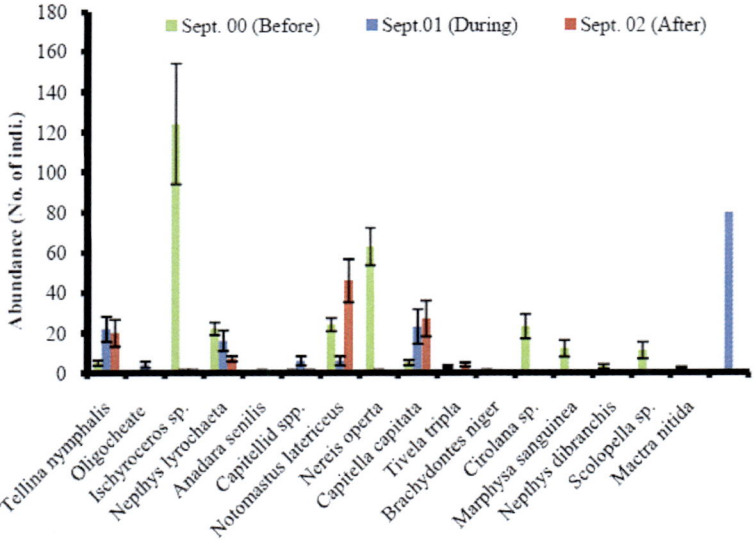

Figure 11. Temporal distribution of key macrobenthic taxa Before, During and After dredging period. Error bars indicate 95% confidence interval.

Table 2. SIMPER analysis results: species contributing to the average Bray–Curtis dissimilarity between the 2000 (Before), 2001 (During) and 2002 (After) dredging based on simultaneous analysis of taxa abundance data. δi: contribution of the i-th faunistic group to the average Bray-Curtis dissimilarity (δ) between the project periods, also expressed as a cumulative percentage ($\sum$δi%). Diss/SD is the ratio of dissimilarity to standard deviation and F is the frequency of occurrence of the 7 sites. For brevity, only species that contributed to ≥ 5.0% and cumulative percentage of ≥70% are listed. The codes in the parenthesis after the species name indicate: 'C' crustacean, 'P' Polychaete, 'B' Bivalve

Species	Ave. Diss	Diss./SD	(δi)	$\sum$δi%)	(F%)
Average dissimilarity between 2000 and 2001 =85.29 %					
Ischyroceros sp. (C)	14.84	1.36	17.40	17.40	38
Nereis operta (P)	13.96	0.97	16.37	33.77	29
Notomastus cf. latericeus (P)	10.20	1.14	11.96	45.72	43
Nepthys lyrochaeta (P)	10.17	1.12	11.93	57.65	52
Tellina nymphalis (B)	8.15	0.75	9.55	67.20	33
Capitella spp. (P)	6.22	0.59	7.41	74.62	24
Average dissimilarity between 2000 and 2002 =83.57 %					
Ischyroceros sp. (C)	15.84	1.29	18.95	18.95	38
Nereis operta (P)	14.55	0.94	17.41	36.36	29
Notomastus cf. latericeus (P)	11.16	1.10	13.36	49.72	43
Nepthys lyrochaeta (P)	9.68	1.00	11.58	61.30	52
Capitella spp. (P)	7.21	0.64	8.62	69.93	24
Tellina nymphalis (B)	5.37	0.76	6.42	76.35	33
Average dissimilarity between 2001 and 2002 =87.71 %					
Nepthys lyrochaeta (P)	23.16	0.73	26.40	26.40	52
Tellina nymphalis (B)	20.51	0.70	23.38	49.79	33
Notomastus cf. latericeus (P)	15.96	0.69	18.20	67.98	43
Capitella spp. (P)	7.33	0.78	8.35	76.34	24
Ischyroceros sp. (C)	6.48	0.46	7.38	83.72	38

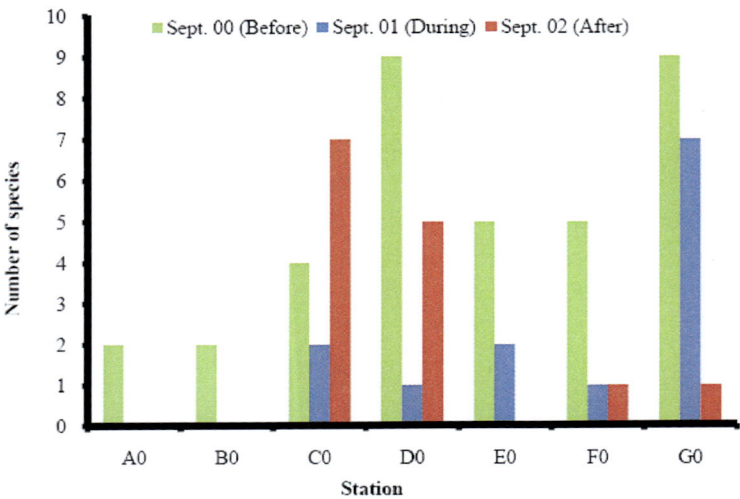

Figure 12. Spatial and temporal distribution of number of species of the dredging periods.

3.4. DREDGING IMPACTS ON WATER QUALITY

The water quality parameters showed both temporal and spatial variation during the periods of the study.

3.4.1. Spatial Pattern of Water Quality

Spatio-temporally, parameters such as total dissolved solids, sulfate and salinity showed a trend. Higher values of these parameters were observed at Stations E-0 and F-0 during May 2002. The lowest values occurred at stations A-0 and D-0. Turbidity did not show any trend, however higher turbidity was recorded at Stations D-0 (Sept. 2001), possibly due to the movement of dredged plume eastwards from Stations F-0 and G-0, which were freshly dredged prior to sampling. Levels of turbidity and suspended solids were extremely low at Station G-0 throughout the study period due possibly to the direction of the fetch carrying the dredged plume eastward (Fig. 13).

The distribution of variables such as dissolved solids, sulfate and salinity seemed to mimic each other spatially. Stations E-0 and F-0 recorded the highest values for these parameters. The other stations recorded values within very narrow ranges.

3.4.2. Temporal Patterns of Water Quality

Temporal distribution of water quality variables in the dredged channel are presented in figures 13 and 14. The surface waters always stayed well oxygenated and generally saturated. The oxygen concentrations between the surface and bottom waters were significantly different throughout the period except in September 2001. There was no existence of dredged channel at September 2000 and hence reference could not be made to the difference between surface and bottom oxygen content during pre-dredging periods. The distribution pattern of the surface and bottom water layers was similar (Fig. 14).

The pH values showed that the dredged channel was moderately basic with slight increase in June 2002. The surface waters were moderately basic than bottom. With a modicum increase of pH from 7.8 in September 2000 to 8.1 in June 2001, the subsequent values remained fairly uniform except the bottom layers which experienced slight variabilities.

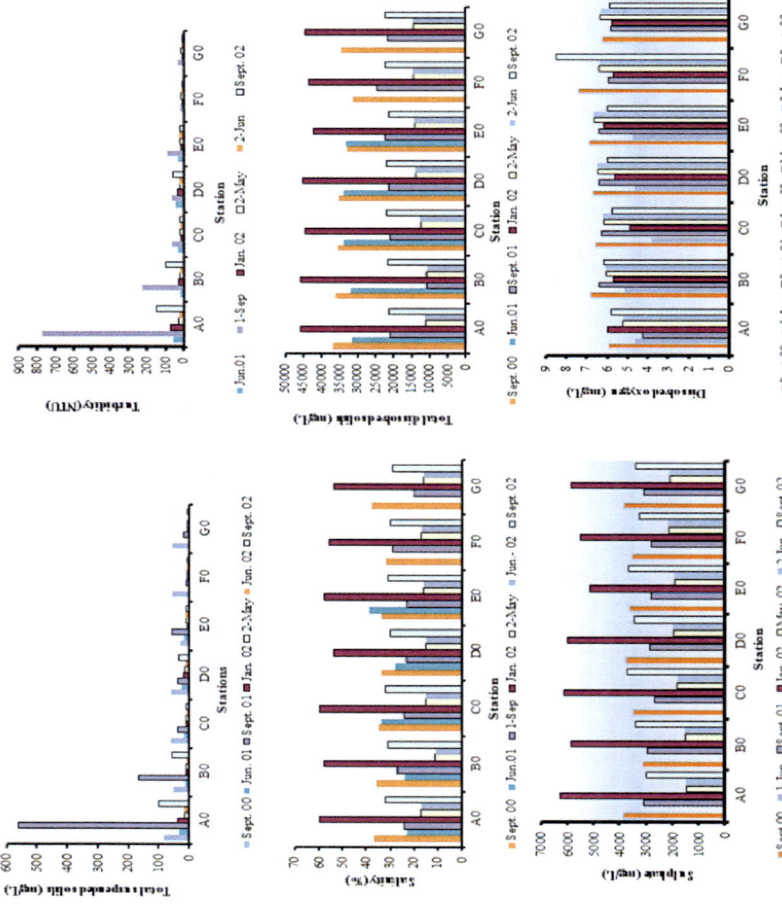

Figure 13a. Spatial distribution of water quality variables at the surface of sampling stations.

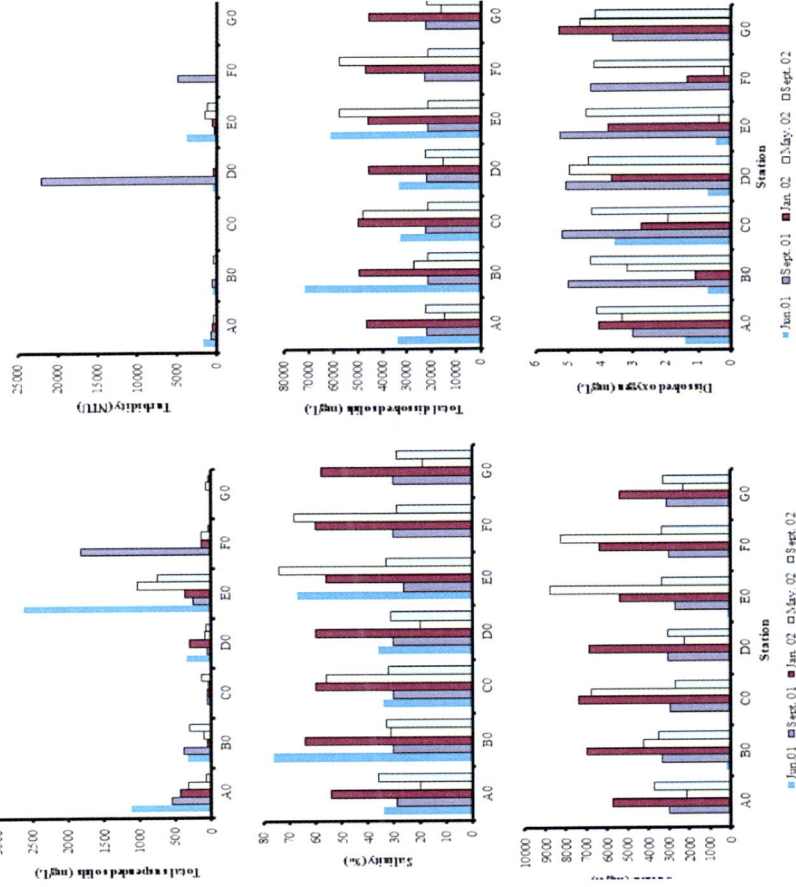

Figure 13b. Spatial distribution of water quality variables at the bottom of sampling stations.

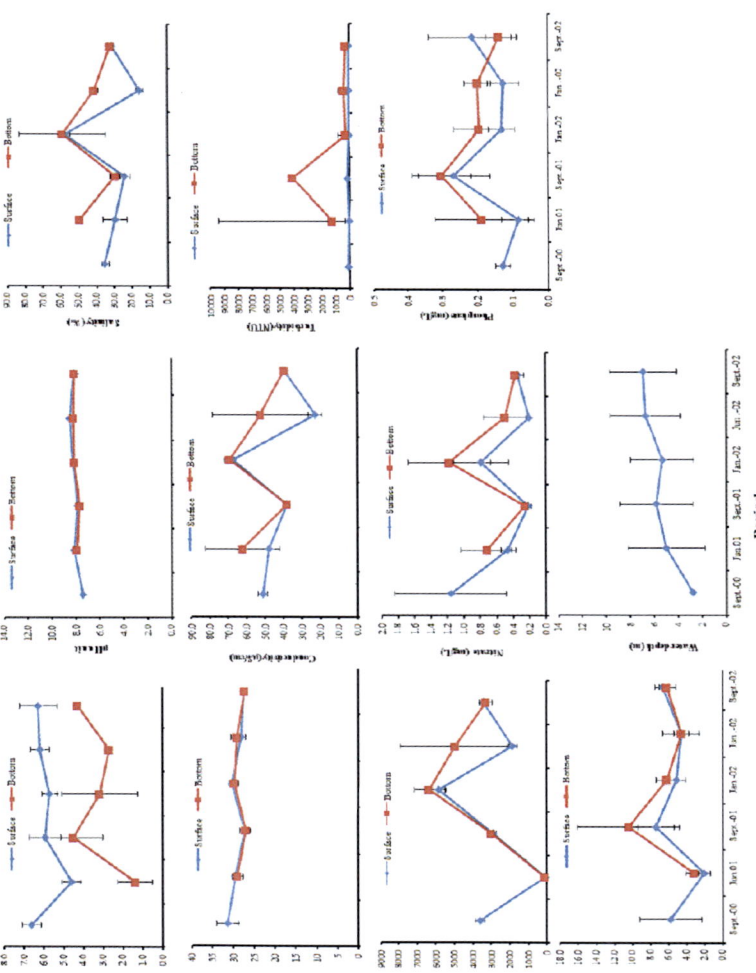

Figure 14. Temporal variations of water quality variables at the surface and bottom of the dredged channel of Keta lagoon.

Water temperatures during the period ranged from 27 to 31°C. Temperatures fluctuated between the periods possibly reflecting the prevailing seasonal climatic conditions. Understandably, surface temperatures were moderately higher than the bottom except in Jun-02 where the contrast occurred. Nonetheless, no significant differences existed between the surface and bottom waters. No abnormal thermal regimes were noticed in the dredged channel. Generally, water temperature influences molecular diffusion and metabolic rates.

The salinity generally mimicked the pattern of water temperature, but showed direct positive relationship with nitrate. Salinity values showed peaks and troughs over the period and between the surface and bottom samples (Fig. 14). The salinity variations realized followed the seasonal patterns with the wet periods recording the lowest. The salinity influences many important processes and functions in aquatic systems. For example, the aggregation and flocculation of suspended particles increase with increasing salinity, meaning that water clarity increases [Håkanson, 2006] leading to increasing primary production of benthic algae and phytoplankton [Preisendorfer, 1986]. It is likely saline water spews from the sediment occurred as a result of dredging operations.

Turbidity is an important parameter which influenced water quality during dredging. Turbidity was generally low except September 2001 during the peak of the dredging when it increased sharply in the bottom waters. Ostensibly, the bottom waters experienced higher turbidity than the surface throughout the period (Fig. 14). The prevalence of moderately higher turbidity at the bottom waters indicates impacts of dredging operations.

Water depth showed moderate increase after January 2002 (After dredging). However, the temporal pattern did not show any significant difference.

Nutrient incursions into coastal lagoons are routed through land drainages and often in sinks/sources. All the nutrients (nitrate, phosphate, silicate and sulfate) measured were generally higher at bottom waters. This indicates the likelihood of bottom sediments releasing quantitative amounts of nutrients into the water column as sequel of the dredging activies. The highest concentrations of phosphate and silicate were recorded in September 2001 during dredging operations (Fig. 14). Frequent resusupension episodes decrease phosphorus content in sediments [Sørensen et al., 2003] resulting in increased concentration in the water column [Watts, 2000]. The importance of sediments as a potential source of phosphorus has also been reported in shallow coastal areas [Fisher et al., 1982]. Anaerobic conditions allow reactions like solubilization of precipitated phosphorus, denitrification

[Day and Yãnez-Arancibia,1985] and therefore always large storage of nitrogen and phosphorus in anaerobic sediments.

Further, the strong positive correlations of salinity and nutrients lend credence to the efflux of saline nutrient-rich water from the sediment due to the dredging. Sediments constitute an essential environmental reservoir in coastal systems due to their capacity to retain and release different compounds from or to the water column.

The result of the water quality index is presented in figure 14. The result generally indicates improved water quality conditions at the surface water layers than the bottom. Using the water quality index scale (Table 3), the result showed impacted water quality by the dredging activities as the quality changed from "Good" before dredging to "Medium" during dredging. Nonetheless, there were improvement after dredging from Jan.-02 to Sept.-02. The WQI index at the bottom layers consistently remained "Medium". However, close examination of the result reveals effects of seasonal influence on water quality as the WQI for Sept.-01 and Sept.-02 at bottom layer were relatively similar (Fig. 14).

Table 3. Water Quality Index scale

WQI	0-25	26-50	51-70	71-90	91-100
Qauality	Very bad	Bad	Medium	Good	Excellent

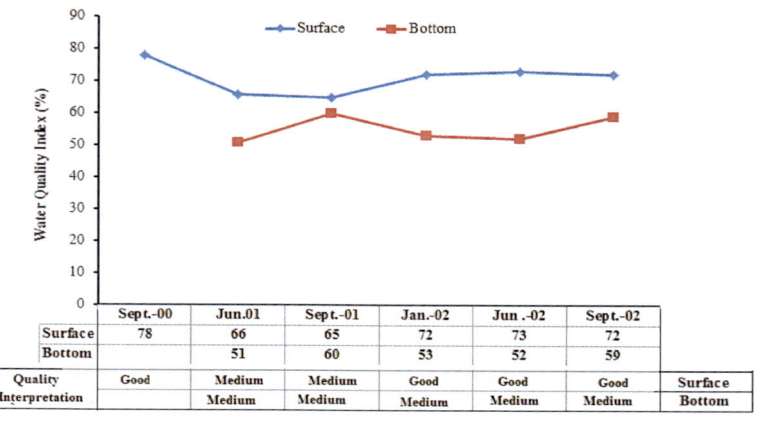

Figure 15. Temporal distribution of water quality index (WQI) for surface and bottom samples.

Chapter 4

CONCLUSION

4.1. IMPACTS OF DREDGING ON SHOREBIRDS COMMUNITY

Shorebirds are an integral aspect of the coastal, wetland, and intertidal ecology of coastal lagoon and wetlands . They contribute to the physical and biological processes of mutual nutrient cycling with direct and indirect effects on the systems and experiencing the effects themselves. The coastal water bodies of Ghana constitute important habitats for both resident and migratory shorebirds from the East Atlantic and the Mediterranean flyways [Ntiamoa-Baidu and Hepburn, 1988; van de Kam et al., 2004]. The shorebirds use these coastal areas for food, resting and breeding . The Keta Lagoon complex is one of the five Ramsar sites under management [Willoughby et al., 2001; Ryan, 2005], with Songor, Sakumo, Densu and Muni-Pomadze are the other four.

Management of these Ramsar sites mainly includes implementation of civil works to minimize the impact of identified problems amongst others. This management objective was considered in the implementation of the restoration project in the study area As such, the dredged materials were used to create habitat islands for the shorebirds. The positive impact of the habitats islands creation on the shorebirds was evident in the increased numerical abundance after the dredging operations (Fig. 7). Notably, the impact was positive for the terns (Fig. 8). This may be attributable to i) the availability/reliable food sources/items such as exposed benthic macro-invertebrates on the habitat island and ii) the creation of the habitat island provided resting, breeding and refugia grounds for the terns notably Little terns, *Sterna albifrons* (Figs. 6 & 7). Rapid exposure of wetland substrate [e.g. Velasquez, 1992] could lead to rapid exposure of fresh patches of unexploited feeding areas which would consequently attract foraging birds. According to Jan *et al.* [2003], foraging waterbirds are uncertain about their chances of success on arrival at a patch and therefore sample their environment to get information on the presence of food so as to 'make decisions'. Since the decision on whether to keep on feeding or leave a patch depends largely on prey encounter rates. The increased abundance of the shorebirds after the project suggests that they found a reliable patch with increased encounter rate as a result of the project.

The waders and 'others' shorebirds categories were negatively impacted by the project as substantial reduction in abundances was realized after the dredging activities in 2002 (Fig. 8). These shorebirds probably moved away from the noise generated by the dredger as a result of the project activities and human presence in the vicinity.

4.2. IMPACT OF DREDGING ON MACROBENTHIC FAUNA

The dredged material (approx. 9091,000 m^3), mainly peat smothered the macrobenthic fauna in area where the sediment were deposited (habitat islands) and also the dredged portion. It however, exposed new area (burrow pit) for colonization. The rate and period of recovery for the macrobenthic fauna constitute important indicators for assessing the enormity of the impacts. Disturbance likely leads to a non-equilibrium state in community structure where communities are continually recovering from the last disturbance [Reice, 1994] as observed during the dredging periods.

In this study, macrobenthic faunal community showed a great deal of dominance and diversity in 2001 (During dredging) and 2002 (After dredging). In fact, during the periods of dredging (2001), the dominant species encountered in the channel was *Tellina nymphalis*, which possibly survived the low oxygen at the bottom of 9 m water depth. This is possibly due to their long siphons, which enable them to filter the water column for food. Gradually, however this species disappeared and could not be recorded in the subsequent samples. Exactly a year later in 2002, colonization by similar species, which were recorded during the dredging and a few months after dredging, showed signs of recovery with appreciable numbers recorded. This probably was due to the more favorable conditions that existed in the channel at the time. It must be noted that the species recorded in 2001 (During dredging) and 2002 (after dredging) were very similar in terms of composition during the wet periods where environmental condition were in the tolerable ranges for the organisms. Lamptey and Armah [2008] indicated higher species diversity and richness of the Keta Lagoon occurred during the wet season. It has also been recognized that coastal soft-bottom macrobenthic communities receive considerable habitat disturbance from natural phenomena, which can result in major changes in salinity , temperature , currents [Flemer et al., 1997], and wave-generated resuspension of sediments [Thistle, 1981; Parker et al. 1980]. Barry and Dayton, [1991] and Thrush et al. [1989] also stressed that, spatial and temporal heterogeneity of lagoons is maintained by a variety of disturbances and other biotic and abiotic factors. This underscores the importance of seasonal abiotic conditions integrating with human-induced impacts in influencing species assemblage patterns and distribution. Generally, benthic fauna community distributions vary considerably in time and space [Boesch, 1973; Carriker, 1967], due, in great part, to the patchiness of species occurrences [Pearson and Rosenberg, 1978] and overall heterogeneity of the benthic habitat [Mistri et al., 2000]. The spatio-temporal heterogeneity has been ascribed to such factors as bottom sediment , spatial variability [Tenore, 1972], climatic irregularity [Bourcier, 1995; Hassle and Sanders, 1967], anthropogenic perturbations [Krönckeet al.,1992; Rosenberg, 1973] and biogenic structures [Woodin, 1981].

In this study, the species abundance showed both spatial and temporal variations . In 2001 (During dredging), the abundance of *Tellina nymphalis* was high at station C-0 and *Capitella capitata* and other Capitellids were high in Station G-0 (where highest species richness and diversity was recorded). This was due to the fact that sampling was done few days after dredging where most of the recorded species were still surviving. However, in 2002 (After dredging), the abundant of species occurred in Stations C-0 (though numbers and diversity varied with the previous year's results) and D-0. In this case, dominant species were *Notomastus* sp. *Capitella capitata* and *Tellina nymphalis* but Station G-0 did not record many species. It could also be realized that the dominant species in each instance were opportunistic species, able to tolerate stressful environments. This corroborates findings by Hall, et al., [1994] that a common consequence of high physical disturbance (subset of processes that lead to the disruption (movement) of sediment) is a numerical reduction in non-opportunistic taxa and increased abundance of opportunistic species.

The possible presence of indicator species , such as the polychaete *Capitella capitata* as well as higher numerical dominance in 2002 (After dredging) may provide additional criteria to monitor the impacts of dredging and rate of colonization . Although *C. capitata* is a complex of sibling species [Grassle and Grassle, 1976], the ecological traits of most species involved are similar [Pearson and Rosenberg, 1978] and the presence of populations of these species represents an index to evaluate the disturbance impact on the communities around the dredged channel, particularly in relation to organically enrichment. The species may also indicate possible rate of re-colonization as in stable communities,the populations of opportunitics species decrease.

However, disturbance of coastal macrobenthic communities induced through experimental manipulation suggest that colonization may be relatively rapid, but time to recovery is variable and depends on the timing of disturbance, nature of the habitat, reproductive periodicity of macrobenthos, and abiotic and biotic factors [e.g. Probert, 1984; Zajac and Whitlatch, 1982]. The results suggest that the re-colonization of the macrobenthic fauna in the Keta lagoon dredged channel may depend on the local hydrodynamic conditions, sediment characteristic and abiotic variables.

The impact of the dredging on the macrobenthic fauna was measurably felt within the few months as the substrates that served as habitats and on which colonization depends were continuously removed and palpably the first 8 months saw no colonization process in place. However, with the emergence of species that were decimated in 2001, a year after (in 2002), the channel is expected to fully recover to stable community in a year or two. This will however largely depend on array of factors including biotic and abiotic.

4.3. IMPACTS OF DREDGING ON WATER QUALITY

Globally, there is increasing awareness that water is one of the most critical natural resources in future. Impacts on water quality is therefore of grave importance particularly of biological resources in aquatic media . It is likely that water sample at any point in time will exhibit varying levels of contaminations with respective to different parameters tested. As such, temporal data from monitoring programs is relevant to elucidate the associated influence on water quality especially in a disturbed aquatic ecosystem.

Temporal and spatial variations were observed in the physico-chemical parameters measured (e.g. salinity, total dissolved solids, total suspended solids and sulfate) in the dredged channel (Figs. 13 & 14). Spatial and temporal variations in turbidity was pronounced. Higher turbidity may have ecological consequences not only adversely affect the production of phytoplankton as it interferes with primary production by limiting light penetration [Jonhston Jn. 1981], but affect fish gills due to its clogging action and can also clog the membranes of filter feeding organisms [Bray 1979].

Many of the water quality variables fluctuated between the periods with a trend that reflect natural seasonal conditions. Ostensibly, estuaries and coastal lagoons are characterized by variability and low predictability of their environmental conditions. Consequently, their physico-chemical parameters exhibit large variation (e.g. seasonal), with the highest concentration generally found following a rainy period. It is therefore sometimes maze decoupling natural phenomenon from human-induced activities such as dredging.

Nonetheless, the water quality index revealed that the water quality was impacted by the dredging activities resulting in a reduction of quality from "Good" (Before dredging) to "Medium" (During dredging). The impact was direct and may be transient as the quality reverted to "Good" from Jan.-02 to Sept.-02. The WQI of the bottom samples indicated the influence of seasonal conditions as Before and After dredging activities, the water quality remained consistently "Medium" (Fig. 15).

This results and findings from this study are relevant to several questions regarding the environmental effects dredging in aquatic ecosystems such as coastal lagoons.

ACKNOWLEDGMENTS

The author would like to acknowledge supports from Environmental Solution Limited (ESL) in Ghana, and Research Planning Inc. (RPI) in the USA.

REFERENCES

A. P. H. A., A. W. W. A., and W. E. F. (1998). Standard methods for the examination of water and wastewater. 20 Washington: American Public Health Association (A.P.H.A.).

Akpati, B. N. (1975). Geological structure and evolution of Keta Basin. Ghana Geological Survey. Report No. 75/3, Ghana.

Alongi, D. M. (1990). The ecology of tropical soft-bottom benthic ecosystems. *Oceanography and Marine Biology Annual Review*, 28, 381–496.

Armah, A. K. (1993). Coastal wetlands of Ghana. *Coastal Zone*, 93, 313-322.

Barr, B. W. (1993). Environmental impacts of small boat navigation: vessel/sediment interactions and management implications. In: Magoon OT, editor. Coastal Zone '93: proceedings of the eighth Symposium on Coastal and Ocean Management; 1993 Jul 19-23; New Orleans, LA. American Shore and Beach Preservation Association. p 1756-1770.

Barry, J. P., and Dayton, P. K. (1991). Physical heterogeneity and the organization of marine communities. In: *Ecological heterogeneity*. Kolasa, J., Pickett, S.T.A. (Eds.) Springer-Verlag, New York, p. 270-320.

Beanish, J., and Jones, B. (2002). Dynamic carbonate sedimentation in a shallow coastal lagoon: Case study of South Sound, Grand Cayman, British West Indies. *Journal of Coastal Research*, 18(2), 254-266.

Begon, M., Harper, J. L., and Townsend, C. R. (1990). *Ecology:* individuals, populations, and communities. Blackwell Scientific Publications, London.

Bemvenuti, C. E., Angonesi, L. G., and Gandra, M. S. (2005). Effects of dredging operations on soft bottom macrofauna in a harbour in the Patos lagoon estuarine region of southern Brazil. *Brazil Journal of Biology*, 65(4), 573-581.

Biney, C. A. (1986). Preliminary physico-chemical studies of lagoons along the Gulf of Guinea in Ghana. *Tropical Ecology*, 27, 147–156.

Bird, E. E. F. (1994). Physical setting and geomorphology of coastal lagoons. In: *Coastal lagoon processes 60 (ed. Kjerfve B.)*, Elsevier Oceanographic Series, Amsterdam. 577pp.

Boesch, D. F. (1973). Classification and community structure of macrobenthos in the Hampton Roads area, Virginia. *Marine Biology* 21, 226-244.

Bonsdorff, E. (1983). Recovery potential of macrozoobenthos from dredging in shallow brackish waters. Fluctuation and succession in marine ecosystems. Proceedings of the 17th European Marine Biology Symposium, Brest, France. *Oceanologica Acta*, 4, 27-32.

Bouncier, M. (1995). Long-term changes (1954 to 1982) in the benthic macrofauna under the combined effects of anthropogenic and climate action (example of one Mediterranean Bay). *Oceanologia Acta*, 19 (I), 67-78.

Bray, R. N. 1979. *Dredging: A Handbook of Engineers*. Edward Arnold Ltd. London, UK. 276pp.

British Geological Survey, (1999). The effective development of offshore aggregate in SE Asia. Technical Report WC/99/9.

Bruun, P. (1962). Sea level rise as a cause of shore erosion. American Society of civil Engineers Proceedings, *Journal of Waterways and Harbors Division*, 88, 117-130.

Bush, D. M., Neal, W. J., Longo, N. J., Lindeman, K. C., Pilkey, D. F., Esteves, L. S., Congleton, J. D., and Pilkey, O. H. (2004). Living with Florida's Atlantic Beaches: Coastal Hazards from Amelia Island to Key West. Durham (NC): Duke University Press.

Carriker. M. R. (1967). Ecology of estuarine benthic invertebrates: a perspective. In: *Estuarine*, ed. G. H. Lauff, pp. 442-487. American Association for Advancement Science Publication, No. 83, Washington DC.

Clark, D. G., and Wilber, D. H. (2000). "Assessment of potential impacts of dredging operations due to sediment resuspension" DOER Technical Notes Collection (ERDC TN-DOER-E9).U.S. Army Engineer Research and Development Center, Vicksburg, M.S.

Clarke, K. R. (1993). Non-parametric multivariate analyses of changes in community structure. *Australian Journal of Ecology*, 18, 117–143.

Clarke, K. R., and Warwick, R. M. (1994). Changes in Marine Communities. An approach to statistical analysis and interpretation. Natural Environment Research Council, U.K. 144pp.

Collins, M. A. (1995). "Dredging-induced near-field resuspended sediment concentrations and source strengths" Miscellaneous Paper D-95-2, U.S. Army Engineer Waterways Experiment Station, Vicksburg, MS.

Dauer, D. M., Ewing, R. M., and Ranasinghe, J. A. (1989). Macrobenthic communities of the lower Chesapeake Bay. Chesapeake Bay Program. Rep. Virg. Water Control board, March 1985-June 1988. Norfolk, Virginia,.

Davoult, D., and Richard, A. (1990). Experimental study of the recruitment of the sessile community on the pebbly se bed in the Dover Strait. *Cah. Bid. Mar.*, 31, 181-199.

Day, J. H. (1967a). A monograph on the polychaeta of Southern Africa. Part 1 Errantia, Trustees of the British Museum London, 458 pp.

Day, J. H. (1967b). A monograph on the polychaeta of Southern Africa. Part II Sedentaria, Trustees of the British Museum London, 877 pp.

Day, J. W., and Yañez-Arancibia, A. (1985). Coastal lagoons and estuaries as an environment for nekton. In Fish community Ecology in Estuaries and coastal lagoons: Towards an ecosystem integration. A. Yañez-Arancibia (Ed.). UNAM Press. pp.17-34.

De Wit, R. (2003). Biodiversity and ecosystem functioning of coastal lagoons. ELOISE Workshop: Demands at the European and Global Level, Goes (The Netherlands) 7-10 May 2003.

Dickson, K. B., and Benneh, G. (1988). *A new geography of Ghana*, London: London Group of companies.

Dominguez, J. M. L. (1984). Sea level history: a dominant control on modern coastal sedimentation style (abstract). Society of Economic Paleontologists and mineralogists, first midyear Meeting, San Jose, California, 26pp.

Edmunds, J. (1978). Sea shells and other molluscs found on West African Coast and Estuaries. Ghana Universities Press, 164 pp.

Emery, K. O., and Stevenson, R. E. (1957). Estuaries and lagoons, physical and chemical characteristics. In: Treatise on Marine Ecology and Paleaecology: Hedgpeth, J.W. (Ed.). *Geological Society of American Memoirs*, 67(1), 673-693.

Entsua-Mensah, M., and Dankwa, H. R. (1997). Traditional knowledge and management of lagoon fisheries in Ghana. Water Research Institute. Technical Report No. 160.

ESL/RPI/GLDD. (2004). Environmental monitoring reports (2001–2004) of the Keta Sea Defence Project Works (KSDPW). Technical Report, Environmental Protection Agency, Ghana.

Espoo, (1991). Convention on Environmental Impact Assessment in a Transboundary Context.

Finlayson, C. M., Gordon, C., Ntiamoa-Baidu, Y., Tumbulto, S. and Storrs. M. (2000). The hydrobiology of Keta and Songor lagoons: Implications for coastal wetland management in Ghana. Supervising Scientist Report 152, Supervising Scientist, Darwin.

Flemer, D. A., Ruth, B. F., Bundrick, C.M., and Gaston, G.R. (1997). Macrobenthic community colonization and community development in dredged material disposal habitats off coastal Louisiana . *Environmental Pollution,* Vol. 96, No. 2, pp.141-154. Elsevier science Ltd. Great Britain.

Fisher, T. R., Carlson, P. R., and R. T. Barber. 1982. Sediment nutrient regeneration in three North Carolina estuaries. *Estuarine Coastal and Shelf Science,* 14, 101-116.

Gaston, G. R. (1985) Effects of hypoxia on macrobenthos of the inner shelf off Cameroon , Louisiana .A review of brine effects and hypoxia. *Gulf Research Reports,* 20, 603-613.

Gaston, G. R. and Edds, K. A. (1994). Long-term study of benthic communities on the continental shelf off Cameroon , Louisiana : a review of brine effects and hypoxia . *Gulf Research Reports,* 9 (1), 57-64.

Grassle, J. P., and Grassle, J. F. (1976). Sibling species in the marine pollution indicator Capitella (Polychaeta). *Science,* 192, 567–569.

Gray, J. S. (1981). The Ecology of Marine Sediments. An Introduction to the Structure and Function of Benthic Communities, First ed. Cambridge University Press, Cambridge. 185 pp.

Green, R. H. (1979). Sampling design and statistical methods for environmental biologists. Wiley and Sons, New York, 257pp.

Goldberg, W. M. (1989). Biological effects of beach restoration in South Florida: The good, the bad, and the ugly. In: Tait, L.S. (ed.), *Beach Preservation technology.* '88: *Problem and Advancements in Beach Nourishment.* Tallahassee: *Florida Shore and Beach Preservation association.* pp.19-27.

Grippo, M. A., Cooper, S., and Massey, A. G. (2007). Effect of beach replenishment projects on waterbird and shorebird communities. *Journal of Coastal Research*, 23 (5), 1088-1096.

Håkanson, L. (2006). Suspended particulate matter in lakes, rivers and marine systems. Caldwell, new Jersey: The Blackburn Press, 331pp.

Hall, S .J., Raffaelli, D., and Thrush, S. F. (1994). Patchiness and disturbance in shallow water benthic assemblages. In: *Aquatic Ecology – Scale, Pattern and Process*, P. S. Giller, A. G. Hildrew, and D. G. Raffaelli, pp. 333-375. Blackwell Scientific Publications, Boston.

Hall, S. L. (1994). Physical disturbance and marine benthic communities: Life in unconsolidated sediments. *Oceanography and Marine Biology Annual Reviews*, 32, 179-239.

Hessle, R. R., and Sanders, H. L. (1967). Faunal diversity in the deep sea. *Deep Sea Research*, 14, 65-78.

Harvey, M., Gauthier, D., and Munro, J. (1998). Temporal changes in the composition and abundance of the macro-bentic invertebrate communities at dredged material disposal sites in the Anse à Beaufils, Baie des Charleurs, eastern Canada. *Marine Pollution Bulletin*, 36(1), 41-55.

Higgs, R. W. (1997). What is good ecological restoration? *Conservation Biology*, 11, 338-348.

Hubbard, D. K. (1992). Hurricane-induced sediment transport in open-shelf tropic systems-an example from St. Croix, U.S. Virgin Island. *Journal of Sedimentary Petrology*, 62(6), 946-960.

Hurlbert, S. H. (1984). Pseudoreplication and the design of ecological field experiments. *Ecological Monograph*, 54, 187-211.

International Council for the Exploration of the Sea. (1992). Report of the ICES working group on the effects of extraction of marine sediments on fisheries. Copenhagen (Denmark): ICES Cooperative Research Report # 182. pp. 877.

International Council for the Exploration of the Sea. (2001). ICES co-operative research report. Report of the ICES Working Group on the effects of extraction of marine sediments on the marine ecosystem. ICES Copenhagen, Denmark, 80 pp.

Jan, A. G., Schenk, I. W., Bos, S., and Piersma, T. (2003). Incompletely informed shorebirds that face digestive constraint maximize net energy gain when exploiting patches. *American Naturalist*, 161, 777–793.

Jensen, A., Mogensen, B., (2000). Effects, ecology and economy. Environmental aspects of dredging - Guide No. 6. International

Association of Dredging Companies (IADC) and Central Dredging Association (CEDA), 119 pp.

Jewett, S. C., Feder, H. M., and Blanchard, A. (1999). Assessment of the benthic environment following offshore placer gold mining in the northeastern Bering Sea. *Marine Environmental Research*, 48, 91-122.

Johnston, Jr., S. A. (1981)."Estuarine dredge and fill activities: A review of impacts". *Environmental Management*, 5, 427-440.

Jumars, P. A., Mayer, L. M., Deming, J. W., Baross, J. A. and Wheatcroft, R. A. (1990). Deep-sea deposit-feeding strategies suggested by environmental and feeding constrains. *Philosophical Transactions of the Royal Society of London. Series* A, 331, 85-101.

Kalbfleisch, W. B. C., and Jones, B. (1998). Sedimentology of shallow, hurricane-affected lagoons: Grand Cayman, British West Indies. *Journal of Coastal Research*, 14, 140-160.

Kapetsky, J. M. (1984). Coastal lagoon fisheries around the worlds: some perspectives on fishery yields and other comparative fishery characteristics. In: *Management of Coastal Lagoon Fisheries,* eds., J.M. Kapesky, and J.M Lasserre, 97-139. FAO Studies and reviews GFCM No. 61. Volume1.

Kates, R. W., Clark, W. C., Corell, R., Hall, J. M., Jaeger, C. C., Lowe, I., McCarthy, J. J., Schellnhuber, H. J., Bolin, B. Dickson, N. M., Faucheux, S., Gallopin, G. C., Grubler, A., Huntley, B., Jager, J., Jodha, N. S., Kasperson, R. E., Mabogunje, A., Matson, P., Mooney, H., Moore, B., III, O'Riordant, T., and Svedin, U. (2001). Sustainability Science. *Science*, 292, 641-642.

Kench, P. S. (1998). A currents of removal approach to interpreting carbonate sedimentary processes. *Marine Geology*, 145, 197-223.

Kenny, A. J. (1995). The biology of marine gravel deposits and the effects commercial dredging. Unpublished PhD thesis. University of East Angila, 243 pp.

Kenny, A. J., Rees, H. L., Greening, J., and Campbell, S. (1998). The effects of gravel extraction of the macrobenthos at an experimental dredge site off North Norfolk, UK (result 3 years post-dredging). *ICES CM* 1998/V, 14, 1-7.

Kjerfve, B. (1994). Coastal lagoons. In: *Coastal lagoon processes (ed. B. Kjerfve),* pp. 1-8. Elsevier Oceanographic Series. Amsterdam.

Kranz, P. M. (1972). The anastrophic burial of bivalves and its paleological significance. PhD thesis, University of Chicago.

Krause, P. R., and McDonnel, K. A. (2000). The Beneficial Reuse of Dredged Material for Upland Disposal. Harding Lawson Associates Engineering and Environmental Services.

Kröncke, I., Duineveld, G.C.A., Raak, S., Rachor, E., and Daan, R. (1992). Effects of a former discharge of drill cuttings on the macrofauna community. *Marine Ecology Progress Series,* 91, 277-287.

Lamptey, E., and Armah, A. K. (2008). Factors Affecting Macrobenthic fauna in a Tropical Hypersaline Coastal Lagoon in Ghana, West Africa. *Journal of Estuaries and Coasts,* 31, 1006-1019.

Lankford, R. R. (1977). Coastal lagoons of Mexico : Their origin and classification . In: Wiley, M.L. (Ed.). *Estuarine Processes,* 2, 182-215.

Levin, L. A. (1984). Life history and dispersal patterns in a dense infaunal polychaete assemblages: community structure and response to disturbance. *Ecology,* 65, 185-200.

Marcovecchio, J., H. Freije, S. De Marco, A. Gavio, L. Ferrer, S. Andrade, O. Beltrame, and R. Asteasuain. (2005). Seasonality of hydrographic variables in a coastal lagoon: Mar Chiquita, Argentina . *Aquatic Conservation, 16,* 335–347.

Martin, L. and Dominguez, J. M. L. (1994). Geological history of coastal lagoons. In:*Coastal lagoon processes* 60. Kjerfve, B. (Ed.). Elsevier Oceanographic series, Amsterdam. 577pp.

Maurer, D., Keck, R., Tinsman, J. C., Leathem, W. A. (1981). Vertical migration and mortality of benthos on dredged material - part 1: Mollusca. *Marine Environmental Research,* 4, 299-319.

Maurer, D., Keck, R., Tinsman, J. C., Leathem, W. A., Wethe, C., Lord, C. and Church, T. M. (1986). Vertical migration and mortality of marine benthos in dredged material: a synthesis . *International Revue der gesamten Hydrobiologie,* 71, 49-63.

Messieh, S. N., Rowell, T. W., Peer, D. L., Cranford, P.J. (1991). The effects of trawling, dredging and ocean dumping on the eastern Canadian continental shelf seabed. *Continental Shelf Research,* 11, 1237-63.

Mistri, M., Fano, E. A., Rossi, G., Caselli, K., and Rossi, R. (2000). Variability in macrobenthos communities in the Valli di Comacchio, northern Italy, an hypereutrophized lagoonal ecosystem. *Estuarine, Coastal and Shelf Science,* 51, 599-611.

Newell, R .C., Seiderer, L .J., and Hitchcock, D. R. (1998).The impact of dredging works in coastal waters: a review of the sensitivity to disturbance and subsequent recovery of biological resources on the seabed, *Oceanography and Marine Biology – An Annual Review,* 36, 127–178.

Nichols, M. M. and Boon III, J. D. (1994). Sediment transport processes in coastal lagoons. In *Coastal lagoon processes 60* Kjerfve, B. (ed.). *Elsevier Oceanography Series.* Amsterdam. 577pp.

Nixon, S.W. (1982). Nutrient dynamics, primary production and fisheries yield of lagoons. In Les Lagunes Côtiére, special volume P. Lassere and H. Postma (ed.), pp 357–371. Océanologia Acta. European Journal of Oceanography Proceedings Symposium.SCOR-IABO-UNESCO, Bordeaux, France.

Ntiamoa-Baidu, Y., and Hepburn, I. R. (1988). Wintering waders in Coastal Ghana. *Royal Society for the Protection of Birds, Conservation Review,* 2, 85–88.

Ormerod, S. J., and Edward, R.W. (1987). The ordination and classification of macro invertebrate assemblages in the catchments of the River Wye in relation to environmental factors. *Freshwater Biology,* 17, 533-546.

Osenberg, C. W. And Schmitt, R. J. (1996). Detecting ecological impacts caused by human activities. In Detecting ecological impacts: concepts and application in coastal habitats. Schmitts, R. J., & Osenberg, C. W. (eds.). Academic Press, London, pp. 3-16.

Pacheco, A. (1984). Seasonal occurrence of finfish and larger invertebrates at three sites in Lower New York Harbor, 1981-1982. Final report. Sandy Hook (NJ): NOAA/NMFS. Special report for USACE, New York District. 53 pp.

Parker, R. H., Crowe, A. L. and Bohme, L. S. (1980) Biological, chemical Survey of Texoma and Capline Sector Salt Dome Brine Disposal Sites off Louisiana, 1978-1979.Vol I. Benthos. National Oceanic and Atmospheric Administration Technical Memorandum, NMFS-SEFC-25. Final report to DOE, NMFS, Southeast Fisheries Center, Galveston, Texas.

Parry, D. M., Kendall, M. A., Rowden, A. A. and Widdicombe, S. (1999). Species body size distribution patterns of marine benthic macrofauna assemblages from contrasting sediment types. *Journal Marine Biological Association* U.K,. 79, 793-801.

Paiva, P.C., 2001. Spatial and temporal variation of a nearshore benthic community in southern Brazil : implications for the design of monitoring programs. *Estuarine, Coastal and Shelf Science.* 52, 423–433.

Pearson, T. H. and Rosenberg, R. (1978) Macrobenthic succession in relation to organic enrichment and pollution of the marine environment. *Oceanography and Marine Biology Annual Review*, 16, 229-311.

Peterson, C. H., and Bishop, M. J. (2005) Assessing the environmental impacts of beach nourishment. *BioScience,* 55, 887-896.

Phleger. F. B. (1969). Some general features of coastal lagoons. In: Ayala-Castañares, A. and F.B. Phleger (Eds.). Coastal lagoons. A symposium. Mem. Symp. Interm. *Coastal Lagoons,* UNAM-UNESCO, Mexico D.F. Nov. 28-30, 1967, pp. 5-26.

Pocklington, P., and Wells, P.G. (1992). Polychaetes. Key taxa for marine environmental quality monitoring. *Marine Pollution Bulletin,* 24, 593–598.

Preisendcrfer, R. W. (1986). Secchi disk science: visual optics of natural waters. *Limnology and Oceanography,* 31, 909-926.

Probert, P. K. (1984). Disturbance, sediment stability, and trophic structure of soft-bottom communities. *Journal of Marine Research,* 42, 893-921.

Rakocinski, C. F., Heard, R. W., LeCroy, S. E., McLelland, J. A. and Simons, T. (1996). Responses by macrobenthic assemblages to extensive beach restoration at Perdido Key, Florida, U.S.A. *Journal of Coastal Research,* 12 (1), 326-353.

Rees, H., Heip, C., Vincx, M., and. Parker, M. M. (1991). Benthic communities: Use in monitoring pointsource discharges . ICES Techniques in Marine Environment Sciences,? 70 pp.

Reice, S. R. (1994). Nonequilibrium determinants of biological community structure. *American Scientist,* 82, 424-435.

Rhoads, D. C., and Boyer, L. F. (1982). The effects of marine benthos on physical properties of sediments : a successional perspective. In Animal-Sediment Relations, McCall, P. L., and Tevez, M. J. S. (eds.), pp. 3-52 Plenum Press, New York.

Rolston III, H. (1994). Environmental ethics : values in and duties to the natural world. Pages 65-84 *in* L. Gruen and D. Jamieson, editors . *Reflecting onnature: readings in environmental philosophy.* Oxford University Press, New York, New York, USA.

Rosenberg, R. (1973). Succession in benthic macrofauna in a Swedish fjord subsequert to the closure of a sulphite pulp mill. *Oikos,* 24, 244-258.

Ryan, J. M. (2005). The Ghana Coastal Wetlands Management Project. http://math.hws.edu/javamath/ryan/Ryan.html

Simenstad, C., Tanner, C., Crandell, C., White, J., and Cordell, J. (2005). Challenges of Habitat Restoration in a Heavily Urbanized Estuary: Evaluating the Investment. *Journal of Coastal Research, Special Issue,* 40, 6-23.

Sørensen, T. H., G. Vølund, A. K. Armah, C. Christiansen, L. B. Jensen, and J. T. Pedersen. (2003). Temporal and spatial variations in concentrations of sediment nutrients and carbon in the Keta lagoon, Ghana. *West African Journal of Applied Ecology*, 4, 89–103.

Steward-Oaten, A., Murdoch, W. M., and Parker, K. R. (1986). Environmental impacts assessment : 'pseudoreplication' in time? *Ecology*, 67, 929-940.

Tenore, K. R. (1972). Macrobenthos of the Pamlico River estuary, North Carolina. *Ecological Monographs*, 42, 51-69.

Thistle, D. (1981). Natural physical disturbances and communities of marine soft bottoms. *Marine Ecology Progress Series*, 6, 223-228.

Thrush, S. F., Hewitt, J. E., and Pridmore, R. D. (1989). Patterns in the spatial arrangements of polycheates and bivalves in intertidal sand flats. *Marine Biology*, 102, 529-535.

Thrush, S. F., Pridmore, R. D. Hewitt, J. E., and Cummings, V. J. (1992). Adult infauna as facilitators of colonization on intertidal sandflats. *Journal of Experimental Marine Biology and Ecology*, 159, 253-265.

Underwood, A. J. (1992). Beyond BACI: The detection of environmental impacts on populations in the real, but variable, world. *Journal of Experimental Marine Biology and Ecology*, 161, 145-178.

Underwood, A. J. (1994). On Beyond BACI: Sampling designs that might reliably detect environmental disturbances . *Ecological Applications*, 4(1), 4-15.

USACE, (2005). Silt curtains as a dredging project management practice. ERDC TN-DOER-E21, September 2005, 18 pp.

USACE, (1983). Dredging and dredged material disposal. Engineering and Design. U.S. Army Corps of Engineers, Engineer Manual EM 1110-2-5025.

van de Kam, J., Ens, B., Piersma, T. & Zwarts, L. (2004) Shorebirds – An Illustrated Behavioural Ecology. KNNV, Netherlands .

Velasquez, C. R. (1992) Managing artificial saltpans as a waterbird habitat: Species' responses to water level manipulation . *Colonial Waterbirds*, 15, 43–55.

Warwick, R. M. (1988). Analysis of community attributes of the macrobenthos of Frierfjord/Langesundfjord at taxonomic levels higher than species . *Marine Ecology Progress Series*, 46, 167–170.

Warwick, R. M., Pearson, T. H., and Ruswahyuni, (1987). Detection of pollution effects on marine macrobenthos: further evaluation of the species abundance/biomass method. *Marine Biology*, 95, 193-200.

Watts, C. J. (2000). The effects of organic matter on sedimentary phosphorus release in an Australian reservoir. *Hydrobiologia*, 431, 13–25.

Williams, B. (1994). Must a concern for the environment be centered on human beings? Pages 46-52 *in* L. Gruen and D. Jamieson, editors . *Reflecting on nature: readings in environmental philosophy.* Oxford University Press, New York, New York, USA.

Willoughby, N., Grimble, R., Ellenbroek, W., Danso, W. & Amatekpor, J. (2001). The wise use of wetlands : identifying development options for Ghana's coastal Ramsar sites. *Hydrobiologia*, 458, 221–234.

Woodin, S. A. (1981). Disturbance and community structure in shallow water sand flat. *Ecology*, 62, 1052-1066.

World Health Organization (WHO) (2004). "Guidelines for Drinking Water Quality," 3rd Edition, Geneva.

Zajac, R. N. and Whitlatch, R. B. (1982). Responses of estuarine infauna to disturbance. I. Spatial and temporal variation of initial recolonisation. *Marine Ecological Progress Series,* 10, 1-14.

Zenkovitch, V. P. (1969). Origin of barrier beaches and lagoon coast, p. 27-28. In Ayala-Castañares, A. and. Phleger F.B. (Eds.). Lagunas Costeras UN Symposia. Mem. Simp. Inetrn. Lagunas Costeras UNAM-UNESCO . Mexico , Nov. 28-30, 1967: 686pp.

INDEX

A

aggregation, 33
algae, 33
aquatic systems, 2, 33
assessment, 6, 7, 51

B

biodiversity, 2
biological processes, 35
biological systems, 6
biomass, 52
biotic, 37, 38
biotic factor, 38
birds, 5, 22, 36
body size, 50
breeding, 35

C

carbon, 51
catchments, 4, 11, 50
causation, 7
clarity, 33
climate, 6, 44
closure, 51
coastal communities, 12
colonization, 7, 8, 9, 36, 38, 46, 52
combined effect, 44
community, vii, 2, 6, 8, 9, 18, 24, 25, 36, 38, 44, 45, 46, 48, 49, 50, 51, 52, 53
complement, 12
composition, vii, viii, 4, 8, 16, 37, 47
compounds, 34
conductivity, 16
conservation, 10
cooling, 5
cost, 7
covering, 16
cumulative percentage, 27
cycling, 35

D

data set, 18
decoupling, 39
degradation, 6
denitrification, 33
deposition, 3, 8, 22
deposits, 3, 4, 48
depression, 3
detection, 7, 52
diffusion, 33
discharges, 51
discrimination, 7, 18
dissolved oxygen, 10, 15, 19
disturbances, 37, 52
diversity, 36, 37, 47

dominance, 36, 38
drainage, 11
drinking water, 18
dumping, 5, 9, 49
durability, 12

E

ecological indicators, 7
ecological restoration, 47
ecology, viii, 1, 15, 35, 43, 47
economy, viii, 1, 47
ecosystem, vii, viii, 2, 5, 7, 12, 38, 45, 47, 49
editors, 51, 52
endangered species, 5
engineering, 12
environmental change, 6
environmental characteristics, 6
environmental conditions, 12, 39
environmental effects, vii, 2, 6, 39
environmental factors, viii, 50
environmental impact, vii, 1, 2, 6, 7, 8, 12, 15, 50, 52
Environmental Protection Agency (EPA), 6, 46
environmental quality, 1, 6
equilibrium, 4, 36
equipment, 10
erosion, 1, 3, 4, 11, 44
ethanol, 16
ethics, 51
evaporation, 16
evapotranspiration, 11
exchange rate, 10
exposure, 36
extraction, 7, 47, 48

F

facilitators, 52
fauna, vii, 6, 8, 9, 12, 15, 18, 26, 36, 37, 38, 49
fidelity, 12
Finland, 8
fish, 5, 10, 39
fisheries, 1, 5, 45, 47, 48, 49
flocculation, 33
flooding, 12
flora, 6, 11
fluctuations, 3, 4, 10
fluid, 4
formaldehyde, 15, 16
formula, 18
freshwater, 11

G

geography, 45
geological history, 4
glycerol, 16
goods and services, 5
grass, 11
Great Lakes, 12

H

habitats, 2, 6, 8, 10, 14, 25, 35, 38, 46, 50
harbors, 5
heterogeneity, 37, 43
human activity, 5
hypothesis, 6
hypoxia, 46

I

Impact Assessment, 6, 46
impacts, vii, viii, 1, 2, 6, 7, 12, 18, 33, 36, 37, 38, 43, 44, 48, 50, 51
indirect effect, 35
integration, 45
intrinsic value, 5
invertebrates, 7, 8, 10, 35, 44, 50

L

lakes, 46
landscape, 6
leakage, 8
lens, 16
light transmittance, 9
Louisiana, 9

M

management, vii, 2, 35, 43, 45, 46, 52
manipulation, 38, 52
marine environment, 50, 51
media, 38
Mediterranean, 35
membranes, 39
migration, 49
mining, 10, 47
mixing, 10
mollusks, 16
monitoring, 2, 6, 7, 15, 38, 46, 50, 51
motivation, 9
multivariate statistics, 18

N

native species, 6
natural resources, 38
nitrate, 15, 16, 19, 33
nitrogen, 34
noise, 36
nutrients, 1, 10, 15, 33, 34, 51

O

ores, 7
organic matter, 10, 52
organism, 5
oxygen, 10, 29, 36
oyster, 5

P

parallel, vii, 3
peat, 3, 25, 36
periodicity, 38
phosphorus, 33, 52
physical environment, 8
physical properties, 51
phytoplankton, 33, 39
pipelines, 8
plants, 8
pollution, 7, 46, 50, 52
positive correlation, 34
positive relationship, 33
power plants, 5
predictability, 39
productivity, 10
project, viii, 1, 2, 5, 6, 7, 11, 12, 16, 18, 22, 27, 35, 36, 52
proportionality, 18
pulp, 51

R

rainfall, 11
reactions, 33
recession, 1
recreation, 5
recycling, 10
regeneration, 46
replacement, 2
replication, 12
resolution, 15
resources, 2, 5, 6, 7, 38, 49
respiration, 10
runoff, 11

S

salinity, viii, 4, 10, 15, 28, 29, 33, 34, 37, 39
sanctuaries, 5
sea level, 1, 3, 4

sea-level rise, 4
sediment, vii, 1, 2, 3, 4, 6, 7, 8, 9, 12, 15, 16, 21, 33, 34, 36, 37, 38, 43, 44, 45, 47, 50, 51
sedimentation, 3, 4, 8, 9, 43, 45
sediments, 1, 4, 8, 9, 10, 21, 22, 33, 37, 47, 51
sensitivity, 2, 49
settlements, 14
shellfish, 7
shoreline, 1, 2
shores, 4, 5, 18
shrimp, 5
shrinkage, 3
sibling, 38
signs, 36
soil erosion, 4
species, viii, 6, 8, 9, 10, 12, 16, 18, 24, 25, 27, 28, 36, 37, 38, 46, 52
species richness, 8, 37
spoil, 9, 12
standard deviation, 27
stars, 7
statistics, 18
storage, 16, 34
streams, 11
substrates, 38
succession, 44, 50
survey, 16
survival, 6
susceptibility, 9
sustainability, 12
sustainable development, 2
synthesis, 49

T

temperature, 4, 10, 15, 33, 37
threats, 11

tides, 4
Togo, 11
tourism, 5
toxicity, 9
traits, 38
transgression, 3
translation, 4
transparency, 15
transport, 4, 8, 47, 49
transport processes, 4, 49
transportation, 1

U

UNESCO, 50, 53
uniform, 29

V

valuation, 2
variations, viii, 2, 32, 33, 37, 39, 51
vegetation, 4
vein, 25

W

waste, 5
wastewater, 43
water quality, vii, viii, 2, 9, 10, 12, 18, 19, 28, 29, 30, 31, 32, 33, 34, 38, 39
West Africa, vii, 10
wetlands, 35, 43, 53
wild animals, 10
wildlife, 2
wind speeds, 11
worms, 7